T0300438

A. Desmond O'Rourke, PhD

The World Apple Market

Pre-publication
REVIEWS,
COMMENTARIES,
EVALUATIONS . . .

"*T*he *World Apple Market* puts into readable form a very complex and rapidly changing global market system. IT IS A 'MUST READ' FOR THOSE INVOLVED WITH FRESH PRODUCE, WHETHER THEY ARE GROWERS, PACKERS, OR MARKETERS. Dr. O'Rourke has collected 30 years of research and the best current information from around the world on the economics of production, consumption, marketing, and trade in apples.

The book provides important information on how the international marketing of apple concentrate affects the returns for local growers of fresh market apples. The author describes the influence of the globalization of the fresh market and its implications for trade and competition in a rapidly changing marketplace.

From the vantage point of 30 years of studying apple marketing, Dr. O'Rourke reflects on future trends and spotlights issues which will confront apple producers. This book provides investors, growers, managers, and government officials with information needed to make decisions based on scientific evidence and data, rather then emotion."

Eugene M. Kupferman, PhD
Extension Horticulturist,
Postharvest Tree Fruit Specialist,
Washington State University

66 **I**n his book, *The World Apple Market*, author Desmond O'Rourke covers many important aspects of the apple industry, including that in the U.S. and throughout the world. Dr. O'Rourke discusses in a useful fashion apple marketing patterns and challenges from grower to consumer, international trade of apples, changing world apple supplies, apple economics regarding production, packing, storage, transportation, retailing, processing, and many other aspects.

THE WORLD APPLE MARKET CONTAINS AN ESPECIALLY USEFUL OVERVIEW OF THE CHANGING AND COMPLEX BUT INCREASINGLY IMPORTANT WORLD SUPPLY OF APPLES AND INTERNATIONAL TRADE ASPECTS. Dr. O'Rourke provides an excellent review of what has happened in the very important world market for apple juice concentrate.

O'Rourke describes with unusual clarity some of the important but complex interrelationships between various segments of the apple industry, including between shippers and retailer-wholesaler buyers and between packer-shippers and growers. He also describes clearly the differences in economic incentives, pricing results, and economic impact at various stages in the apple marketing system.

The World Apple Market provides an excellent overview of the apple industry in Washington, the United States, and the world. Dr. O'Rourke is to be commended in attempting such a complex undertaking regarding this important and growing industry."

Donald Ricks, PhD
Professor of Agricultural Economics,
Michigan State University

The World Apple Market

FPP Agricultural Commodity Economics,
Distribution, & Marketing
A. Desmond O'Rourke, PhD

New, Recent, and Forthcoming Titles:

Marketing Livestock and Meat by William Lesser

Understanding the Japanese Food and Agrimarket: A Multifaceted Opportunity edited by A. Desmond O'Rourke

The World Apple Market by A. Desmond O'Rourke

Marketing Beef in Japan by William A. Kerr et al.

The World Apple Market

A. Desmond O'Rourke, PhD

CRC Press
Taylor & Francis Group
Boca Raton   London   New York

CRC Press is an imprint of the
Taylor & Francis Group, an informa business

Reprinted 2009 by CRC Press

Library of Congress Cataloging-in-Publication Data

O'Rourke, A. Desmond (Andrew Desmond), 1938–
    The world apple market / A. Desmond O'Rourke.
        p.   cm.
    Includes bibliographical references and index.
    ISBN 1-56022-041-4 (acid-free paper).
    1. Apple industry. I. Title
HD9259.A5074   1993
382'.41411–dc20   93-7919
                                                                                                                CIP

ABOUT THE AUTHOR

Andrew Desmond O'Rourke, PhD, is Professor of Agricultural Economics, and Director of the International Marketing Program for Agricultural Commodities and Trade at Washington State University in Pullman, Washington. Dr. O'Rourke has been involved in research, teaching, and consulting in the domestic and international aspects of marketing since 1960 in both the public and private sectors. He is the author of two textbooks, a contributor to a number of other books, and the author or co-author of over 150 journal articles, research bulletins, and popular articles on marketing topics. His areas of interest have included marketing of North American fruits, beef, dairy products, seafood, vegetables, and grains in all five continents. Since 1985, Dr. O'Rourke has been Director of the IMPACT Center, an integral part of the College of Agriculture and Home Economics at Washington State University. The IMPACT Center is a multidisciplinary program which applies science and technology to the solution of international agricultural marketing problems.

CONTENTS

List of Tables

List of Figures

Foreword

The World Apple Market gives a comprehensive view of the apple industry. With the changing worldwide distribution of fresh apples and apple products, the reader can view both regional and world developments. This overview of the global apple industry is a masterful effort by Dr. Desmond O'Rourke to bring together the many functions of a highly competitive industry.

Historical background as well as current operation and future possibilities are covered for production, marketing systems, transportation, and government involvement. A majority of the major studies involving apples have been used and effectively integrated into this book.

For any person thinking of entering the apple industry, *The World Apple Market* provides an excellent background at any level. O'Rourke, with his vast experience in the economic factors of the apple industry, has included growing, selling, processing, transportation, and the management functions of these processes. It is also of potential value to anyone interested in the innumerable functions involved in an industry characterized by near "pure competition" theory of economics. The less-than-perfect gathering and sharing of information and extensive regulation by many governments interfere with the performance of pure competition.

Industry and agency efforts to provide information and some of the possible future improvements are suggested. Also the major current activities to overcome government regulations are reviewed.

Some of the changes in the apple industry that are now occurring, as well as some in the future, are discussed. Adapting to these changes will be crucial to most people involved in the many functions of the apple industry. The producer, who receives what is left of the retail price after all other functions are paid for, is the most affected by these changes. The producer should be the most aware of current as well as future changes.

The World Apple Market brings into focus the vast and complicated growing and marketing of the world apple system. O'Rourke, in his ability to convey his message with clearness and simplicity, has provided a book that can be comprehended by novice and expert alike. Through its extensive list of references, the reader can pursue complicated subjects in greater depth. This book is a superior reference encompassing virtually all phases of the apple industry and will be welcomed by all concerned.

Richard D. Bartram
Marketing Consultant

Introduction

Apples are among the oldest food products known to man. The apple figures prominently in the Biblical story of Adam and Eve. It is also featured in the earliest records of Greek civilization. Long important in Europe and Asia, apples were brought to and flourished in suitable locations in many colonies in temperate climates around the world. They have grown steadily in the twentieth century in total production, consumption and trade.

Like many other agricultural commodities, however, apples have undergone a dramatic transformation in the last quarter of the twentieth century. Growers, shippers and traders have had to deal with the macroeconomic changes that have convulsed the international economic system. These included the breakdown in 1971 of the fixed exchange rate system set up at Bretton Woods in 1945, the explosion in oil prices after the Yom Kippur War of 1972, the recycling of petrodollars which stimulated the growth of world trade in the 1970s, and the resultant inflation, rise in interest rates and debt crises which assailed the developing countries through the 1980s and into the 1990s.

During the same period, the rapid expansion of production combined with increasing availability of competing fruits to put downward pressure on prices and profitability. The more progressive producers and producing regions have aggressively adopted changes in production, handling, storage, processing, marketing and transportation. This has raised the competitive stakes for those regions who wish to remain in the game.

An industry which was based on natural climatic advantage has been transformed into an industry where investment in new technology has increasingly become the key to competitiveness. An industry with its practices steeped in long cultural traditions has had to adopt radical departures in financing, in planting systems, in marketing organizations and in the speed with which its systems are renewed.

1

Technology has also enlarged the number of acres in row crops that can be shifted into apple production if investment is available. Thus, new regions can be expected to emerge to challenge the existing dominant producing regions. At the same time, supplies of competing fruits such as oranges, bananas, kiwi fruit and nectarines are likely to keep the competitive pressure on apples. As it enters the twenty-first century, the world apple industry will increasingly become dominated by those regions that have the managerial and financial resources and technical acumen to adapt.

This book is intended to analyze the existing situation in the world apple system from production, through assembly, packing, processing, storage, transportation, distribution, marketing and consumption. It examines how the forces of globalization have affected the industry, the managerial and technological adaptations that have been adopted to deal with changing situations, and the economics faced by participants in the industry as they attempt to prepare for the twenty-first century.

Chapter 1

World Apple Supplies

Most of the world's supplies of apples come from the temperate zone of the Northern and Southern hemispheres between latitudes 40° and 50° north in Europe and North America, 30° and 40° north in Asia and 20° and 40° south in the Southern Hemisphere. Historically, production has been possible outside these zones where climate was moderated by ocean influences (as in New Zealand) or by higher elevations in some semitropical areas.

Apples require sufficient cold temperatures in winter to induce dormancy and the setting of fruit in the subsequent season. They require long, warm summers in order to obtain ripening of most varieties. Thus, apples do not generally thrive in colder, continental climates. However, technology has permitted a gradual expansion of the range of areas in which apples can be grown successfully. Advances in cold hardiness of rootstock and in frost control have permitted production in harsher winter and spring conditions. The development of irrigation and of more heat-tolerant varieties has permitted expansion into semitropical desert conditions. Over time, the industry appears to be inching closer to the equator as the advantages of climate control in the warmer areas offset the greater hazards of production in the cooler climates.

PRODUCTION TRENDS

World apple production grew steadily in the decades of the 1960s, 1970s and 1980s (Table 1.1). This was similar to the experience of deciduous fruits such as pears, peaches, apricots, and cherries. However, while apples became relatively more important

3

TABLE 1.1. World Production of Major Deciduous Fruit, Selected Years

	Annual Average Production (1000 metric tons)				
	1948-62	1961-65	1969-71	1979-81	1986-88
FRUIT					
Apples	13,512	18,175	28,309	34,551	40,114
Pears	3,926	5,552	7,923	8,559	9,633
Peaches	2,443	4,765	6,289	7,254	7,938
Apricots	707	1,070	1,524	1,700	2,001
Cherries	1,214	1,635	1,570	1,178	n/a
TOTAL	21,802	31,197	45,615	52,982	n/a

SOURCE: United Nations Food and Agriculture Organization. *Production Yearbook,* Annual (selected issues).

among deciduous fruits, they were facing increasing supplies of tropical and semitropical fruits and of more exotic items such as kiwi fruit, avocados, mangos, and papayas. Per capita production of apples and other deciduous fruits grew little in the 1970s and 1980s.

Expansion of apple production has not been uniform around the world (Table 1.2). While the quality of statistics on apples varies by country, we can only assume that the annual production figures reported by the United Nations Food and Agriculture Organization at least correctly capture the direction of trends in production. Throughout the period studied, the European continent has remained the dominant supplier of apples. Growth in western Europe was rapid in the 1960s but tapered off in the 1970s as the European Community began a grubbing program to offset an oversupply problem. During the same period, production was reported to have expanded rapidly in the centrally planned countries of Europe and Asia, although these statistics cannot be easily verified.

While some growth occurred in North America, it was particularly rapid in the Southern Hemisphere and in nontraditional suppliers such as India, Iran, Pakistan and Turkey. Much production expansion in the 1970s was stimulated by generous credit from agencies such as the World Bank, individual governments interested in development of intensive agriculture and private investors.

While investment in new plantings slowed dramatically during the 1980s, production continued to increase in many areas as the new plantings reached full maturity. Growth was greatest in the United States, Mexico, Chile, New Zealand, South Africa, Brazil, Turkey, and France. Expansion of plantings continued in Southern Hemisphere countries such as Chile, New Zealand, and Brazil. Orchard abandonment increased in more marginal production areas in Europe, North America and Australia. Political and economic turmoil dramatically altered the environment for production and marketing of apples in many centrally planned countries.

The outlook for production in the 1990s reflected a continuation of these trends. Total European production was expected to stagnate. However, growth was expected to continue in the United States, Mexico, selected countries of south and east Asia, and among leading producers in the Southern Hemisphere such as Chile, Brazil, South Africa, Australia, and New Zealand. Considerable uncertainty surrounded prospects for the former USSR and the People's Republic of China. It seemed unlikely, however, that world apple production would grow faster than world population in the 1990s.

APPLE VARIETIES

While apple trees frequently remain in production even in commercial orchards for many decades, there is a continual process of abandoning older varieties and strains as newer varieties are tried and prove successful. There is also considerable seesawing in the popularity of varieties from region to region. For example, during the 1970s, as Granny Smith plantings were expanding in the Northern Hemisphere, they were contracting in the Southern Hemisphere (O'Rourke, 1986b).

Varieties can be distinguished in many ways (Norton and King, 1987). For example, varieties vary in their color, usually variants of red, yellow or green. The color may be solid, striped, or have a

TABLE 1.2. World Production of Apples by Continent and Major Producing Countries, Selected Years

Region or Country	Annual Average Production (1000 metric tons)				
	1948-52	1961-65	1969-71	1979-81	1989-91
World	13,512	18,175	28,309	34,503	40,181
Europe, total	9,508	11,822	13,966	13,359	14,148
European Community	7,093	8,384	7,531	6,877	7,404
Eastern Europe	1,117	1,806	4,677	4,744	5,013
Scandinavia	213	221	211	205	129
Other	1,085	1,410	1,547	2,133	1,602
North America, total	2,758	3,208	3,550	4,457	5,486
Canada	300	416	413	466	513
Mexico	49	108	175	280	548
U.S.A.	2,409	2,684	2,962	3,711	4,425
South America, total	273	536	718	1,406	2,189
Argentina	195	440	435	945	1,006
Chile	45	54	142	252	683
Other	33	42	140	209	(500)

Africa, total	61	173	270	460	645
South Africa	39	127	224	375	545
Other	22	46	46	85	(100)
Oceania, total	247	426	556	533	718
Australia	199	340	430	328	325
New Zealand	48	86	126	205	393
USSR (former)	—	—	4,533	6,445	6,300
Asia, total	676	2002	4,710	7,963	10,695
China	118	305	1,953	2,993	(3,967)
India	n.a.	n.a.	277	734	(881)
Iran	n.a.	n.a.	89	460	(1,022)
Iraq	n.a.	n.a.	36	103	(72)
Japan	382	1,066	1,037	887	1,048
Korea, Rep.	48	120	217	459	(601)
Korea, DPR	n.a.	n.a.	135	460	(618)
Lebanon	14	96	98	122	(78)
Pakistan	3	13	34	100	(208)
Turkey	102	326	716	1,419	1,900
Other	9	76	107	226	(300)

SOURCE: United Nations Food and Agriculture Organization. *Production Yearbook*, Annual (selected issues) (Numbers in parentheses are author's estimates.)

blush of a different hue. They vary in shape: round, conic, oblate, or oblong when looked at from the front; regular, elliptical, angular, or ribbed when looked at from above. They vary in taste as a result of skin toughness, flesh texture, sweetness, or juiciness.

Varieties also vary in their adaptability to differing climatic conditions, soil types, and rootstocks; in their ease of management, in their susceptibility to disease, winter injury, drought and rainfall; and in the quantity, quality and consistency of yield. They vary in their durability in packing, storage and shipment and in their appropriateness for different end uses such as eating out of hand, home cooking, food processing (e.g., as juice), or in industrial uses.

Clearly, the long-term changes in the popularity of different varieties are affected by changes in the technology of growing, handling, processing and marketing apples and by changes in consumer preferences. While economists can identify the supply response of individual varieties to price factors, there has never been a comprehensive study of the multiple factors affecting changes in varietal preference among growers, marketers and consumers. Thus, there is no simple model to help predict what sort of varietal changes we may expect in the world market in the future. Only fragmentary evidence is available on world production of major varieties. The USDA's Foreign Agricultural Service, Horticultural and Tropical Products Division developed an estimate of apple production by varieties for 32 countries in 1977 and 1978. While the People's Republic of China and the former USSR were not included, the survey covered over 90 percent of the remaining production recorded by UNFAO for those years.

In 1977-78, the leading variety worldwide was the Delicious, with 27.6 percent of production in the 32 countries, followed closely by Golden Delicious with 22.8 percent. Much smaller shares were accounted for by Jonathan (3.3 percent), McIntosh (4.0 percent), Granny Smith (4.4 percent) and Rome Beauty (4.9 percent). Golden Delicious production was significant enough to be reported separately for 23 countries, Delicious for 21 countries, and Jonathan for 17 countries. Golden Delicious was most important in Europe, Delicious outside Europe in the Americas, Asia and Oceania. Jonathan was most important in Eastern Europe and Australia. Granny Smith volume exceeded 20 percent of production only in the Southern Hemisphere countries of Argentina, Australia, Chile, New Zealand and South Africa, although substantial production was reported for France, Italy and Spain. McIntosh were important only in North America and Poland, Rome Beauty in Italy and Cox's Orange in Northern Europe.

The USDA (FAS) attempted a follow-up world apple survey a decade later but was only able to obtain data for 15 countries covered in the 1977-78 survey, and for mainland China. As reported by Castaldi (1988), Delicious and Golden Delicious had measurable production in all 16 countries. Delicious share of production had risen 13 percentage points in Hungary and fallen 42 percentage points in Japan in the decade. Golden Delicious had lost over 14 percentage points in the Netherlands, while Granny Smith had gained market share in Chile, France and New Zealand and lost ground in Argentina and South Africa.

To further examine changes in varietal importance over time, we attempted to piece together from various sources variety data for major apple producing countries in the 1989 and 1990 seasons (Table 1.3). Failing that, we used available variety data for the most recent two seasons. For the 17 countries for which comparable data were available, total apple production rose by 21.7 percent. All but four countries showed gains in total production.

Of the leading varieties, Delicious and Golden Delicious showed a small increase in volume and a small decline in market share. Jonathan, Rome Beauty, and McIntosh lost both volume and market share. Granny Smith and all other varieties gained both volume and market share.

Data on the makeup of all other varieties is incomplete. However, different minor varieties are of considerable importance in different countries (Table 1.4). Traditional varieties such as Boskop, Idared, James Grieve and Ingrid Marie remain important in Europe, where newer varieties, such as Jonagold and Elstar have been increasingly important. Varieties such as Fuji and Mutsu, which first came to prominence in Japan, and various strains of Gala, first developed in New Zealand, have now spread to other countries and are likely to become increasingly important in the 1990s. Still other varieties, including Ralls in China, Amasya in Turkey, Tsugaru and Ohrin in

TABLE 1.3. Production of Apples by Major Varieties, Selected Countries,[1] Annual Average, 1977-78 and 1989-90 (metric tons and percent)

Variety	Volume		Share	
	1977-78	1989-90	1977-78	1989-90
	(1000 metric tons)		(Percent)	
Delicious	3,530.4	3,625.0	27.6	23.3
Golden Delicious	3,311.2	3,834.0	25.9	24.6
Jonathan	421.1	179.0	3.3	1.1
Granny Smith	559.9	1,385.0	4.4	8.9
Rome Beauty	628.3	457.0	4.9	2.9
Cox's Orange	277.6	453.0	2.2	2.9
McIntosh	510.5	466.0	4.0	3.0
All Other	3,567.0	5,186.0	27.9	33.3
TOTAL	12,806.0	15,585.0	100.0	100.0

SOURCE:

1977-78: USDA, FAS. *Apple Production by Major Varieties for Selected Countries.* Horticultural and Tropical Products Division, Washington, DC, 1979.

1989-90: USDA, FAS. Fresh Deciduous Fruit Reports. (Unpublished, occasional attaché reports.)

[1]Countries included Argentina, Belgium, Canada, Chile, France, W. Germany, Greece, Italy, Japan, Netherlands, New Zealand, S. Africa, Spain, Switzerland, United Kingdom, and United States.

Japan, and Bramley in the United Kingdom, are important in major producing countries but little known or appreciated elsewhere.

Information on varietal changes over time is limited. Between the mid-1970s and the late 1980s, Delicious production more than doubled in Chile, New Zealand and South Africa and increased by 73 percent in the United States, but fell precipitously in Japan.

Golden Delicious production continued to grow in most of Western Europe and in the United States. Granny Smith production enjoyed large increases in the Southern Hemisphere, France, Italy, and the United States. Production of Cox's Orange rose sharply in West Germany and the U.K. McIntosh, Rome Beauty, and Jonathan lost ground in most major supply regions.

In the 1990s, one can expect much greater experimentation by all producing regions with varieties that are relatively new or previously unsuccessful. With increasing market pressures and new technologies, producers will alter their choice of variety, sometimes in unpredictable ways. Delicious, Golden Delicious, and Granny Smith can be expected to lose market share to varieties such as Braeburn, Elstar, Fuji, Gala, and Jonagold. Depending on consumer acceptance, price and ease of production and handling, one or more of these varieties may experience in the 1990s the sort of market share growth experienced by Granny Smiths in the 1980s. Correct choice of variety will be increasingly important in competing in world markets.

CONSUMPTION AND USE

Until modern times, apples were generally used or consumed close to the point of production. Thus, the choice available to consumers was determined by the planting and harvesting decisions of local growers. Because apple orchards can remain physically productive for decades, traditional patterns of availability tend to have a persistent effect on consumption. In any market, one can find varieties with only local appeal selling side by side with fruit from distant states, countries and even a different hemisphere.

Unfortunately, little reliable data exists on how consumers in different countries use different apple varieties. In the United Kingdom, apples were traditionally classified as dessert apples or cooking apples. Dessert apples, such as Cox's Orange, would be bought in the retail fruit store or grocery store for eating out of hand at meals and snacks. Cooking apples, such as Bramley, would be bought in the same outlets but prepared at home into pies, tarts or cakes. A small share of the crop was manufactured into alcoholic

TABLE 1.4. Production of Apples by Variety by Country, 1989-90 (1000 metric tons)

Variety

Country	Delicious	Golden Delicious	Granny Smith	Jonagold	Boskop	Fuji	Gala	Cox's Orange
Argentina[2]	119	121	247					
Belgium		62		117	28			15
Canada[3]	126							
Chile	300		180					
France	183	1184	235					
Germany, W.[4]		251		89	37			205
Greece	171	31	13					
Italy	366	862	60	34				
Japan	88	8		54		537		
Netherlands		90		70	35			60
N. Zealand[1]	67	18	111				19	24
S. Africa[2]	119	121	247					
Spain	140	335						
Sweden								5
Switzerland[4]		52		11				
U.K.[4]								144
United States	1946	699	292					
TOTAL	3625	3834	1385	375	100	537	19	453

SOURCE: USDA, FAS. Fresh Deciduous Fruit Reports. (Unpublished occasional attaché reports.)
[1]1984   [2]1984-85   [3]1987-88   [4]1988-89

Variety

Rome Beauty	Gloster	Elstar	Mutsu	James Grieve	Ingrid Marie	McIntosh	Ida-Red	Jonathan	Other	TOTAL
24									26	537
		8	3	4					48	285
						183			187	496
									—	480
									222	1824
				38	75		83		1008	1786
									63	278
180									406	1908
			30					17	323	1057
		64							85	424
				20				3	44	286
									50	537
									214	689
									14	28
	5				9		17		70	155
253									210	354
						283	73	159	756	4461
457	5	72	33	62	84	466	173	179	3726	15585

cider. However, early in the twentieth century the United Kingdom began to import fresh dessert apples in the off season from its Southern Hemisphere colonies like Australia, New Zealand, and South Africa, and these volumes increased steadily after World War II. Even before its entry into the European Community in 1973, and increasingly since, French Golden Delicious has become the leading dessert variety sold in the United Kingdom.

West Germany was even more dependent on local suppliers. Bagley (1977) reported that because of the abundance of noncommercial (pasture) orchards, individual producers in 1973 still consumed 39 percent of the harvested production of apples in West Germany, a decline from the 51 percent share consumed in 1958-62. Even in the late 1980s, it was estimated that only one-third of total West German apple production came from commercial (intensive) orchards.

In contrast, the French apple industry went through a remarkable transformation in the late 1950s and early 1960s. Between 1948 and 1957, France's annual apple production averaged over four million metric tons, about twice that of the next largest producer, the United States. About 90 percent of French apple production was used either for cider or for industrial purposes. After 1958, increased apple plantings in southern France of dessert apples, especially Golden Delicious, and grubbing of older orchards, made France predominantly a producer of dessert apples. Italy, the other major producer in Western Europe, has been slower than France but more successful than West Germany in converting to commercial orchards producing dessert apples.

The United States shared the European experience of mixed commercial and backyard apple plantings prior to World War II. Particularly in the Northeast, many varieties had local or regional popularity. However, in the 1950s, the industry became dominated by commercial producers. Two seemingly contradictory trends emerged. Apples shared in the economywide shift towards processed convenience foods. Products such as applesauce, canned apples, and apple juice became staples on supermarket shelves. The production of dried and frozen apples for domestic and commercial baking use expanded. At the same time, the proportion of apple production accounted for by dessert apples such as Red Delicious,

Golden Delicious, and McIntosh grew to almost two-thirds of total supplies. One explanation for this apparent anomaly was that Golden Delicious was innately dual purpose in use, while advances in food technology permitted acceptable juices to be made out of the leading dessert variety, Red Delicious.

After the mid-1970s, demand in the U.S. for all processed convenience foods, except apple juice, stagnated while per capita consumption of fresh fruit, including apples, rose. The apple juice industry was revolutionized by the availability of supplies of apple juice concentrate from all over the world. Apple cider, which had been largely a local, seasonal product, has become an increasingly widely distributed, commercial, manufactured product.

However, apple varieties in the U.S. continue to be distinguished by purpose. The leading fresh varieties are Red Delicious, McIntosh, Winesap, Golden Delicious (from the western states), and Granny Smith. Virtually all of the new apple varieties being promoted are also in the fresh category. A small number of varieties, York, R.I. Greening, Gravenstein (in California), and Northern Spy (in Michigan) are still grown primarily for processing. A third category is called "dual purpose" because a significant demand exists in both the fresh and processing markets. This category includes Jonathan, Golden Delicious (from the Eastern and Central U.S.), Rome, Stayman, Newton, Cortland, and Idared. Given the fact that more offgrade Red Delicious are processed into juice than the total production of many so-called processing varieties, and that Golden Delicious can be classified as a fresh variety in one region and a dual purpose variety in another, these categories are clearly rather arbitrary.

Given these differences in how apples are classified in different countries, comparisons of supply or use by categories between countries becomes impossible. The USDA makes a herculean effort to develop such comparisons using the simplest possible categories.

Supply in any year is assumed to be equal to domestic production plus imports during the crop year. Stocks are assumed to be zero at the end of one crop year and the beginning of the next crop year. In fact, some fresh apples from long-term cold atmosphere storage may be carried over from one crop season to the next. However, the amount is very small and not likely to distort statistical comparisons between years. A more serious problem arises with processed apples such as canned, frozen, dried or juice products, where carryover can vary widely from year to year. In the absence of adequate data, one can only ignore the problem and proceed with caution.

Supply (as defined above) is then divided into the quantity consumed fresh, fresh exports, processing use and withdrawals. Fresh use might be more aptly called "disappearance" since there will be losses in storage, transportation and distribution from fruit intended for the fresh market due to damage, decay, shrinkage, etc. The USDA assumes that about four percent by weight of the apples shipped to fresh market in the United States will be lost before they are purchased by consumers at retail. Further losses may occur during home storage. Losses will tend to be greater in market channels where handling before shipment is inadequate, transportation conditions of vehicles, highways or railroads are difficult, weather conditions are excessively hot and humid or cold, or refrigerated storage or climate-controlled storage or transportation facilities are limited.

Data on fresh exports or imports are relatively reliable and are incorporated in the USDA balance sheets, but exports and imports of processed products are not. Most governments have a strong incentive to monitor shipments across their borders for the purposes of collecting tariff revenues, preventing the introduction of plant diseases or pests, or buttressing domestic aid programs to farmers. In addition, the statistics of the exporting country can be compared against those of the importing country. Despite these advantages, trade data can have wide errors due to smuggling, unrecorded reexports (especially through entrepot ports such as Rotterdam and Singapore) and mixed cargos.

Processing use tends to be quite difficult to measure. Unlike production, processing plants tend to be located both in major producing areas and in urban areas near to large concentrations of consumers. Plant output of finished product is most easily classified and counted, but it is difficult to aggregate gallons of juice with pounds of dehydrated product and cans of applesauce. Conversion ratios between raw material weight and finished apple product vary by variety and season. However, for the purpose of its apple balance sheet, the USDA reports processing use in terms of estimated raw tonnage of apples used.

The last category reported by the USDA is withdrawals from market. Withdrawals tend to occur primarily in Europe where countries frequently struggle with surpluses of less desirable apples. Withdrawals may include sales into official storage as in the European Community's "Intervention" program, conversion into alcohol as in Switzerland, sales for animal feed, or sales in other forms that render the product unsalable for human food.

Despite all these inadequacies, the USDA balance sheet approach does give a bird's-eye view of where major markets get their supplies of apples and how they utilize those supplies (Table 1.5). Average data for the crop years 1989 and 1990 are presented for most of the major apple producing countries in Europe, North America and the Southern Hemisphere.

The United States produced twice the volume of its nearest competitor for which data were available. France, West Germany, Italy, and Turkey were next with production close to two million metric tons. Hungary, Japan and Argentina were in a third grouping with production close to one million metric tons. Spain, Yugoslavia, Canada, Mexico, Chile, New Zealand, and South Africa were in the next rank with 400 to 700 thousand metric tons.

West Germany and the United Kingdom were the largest importers in volume terms. While imports reached or exceeded 100,000 metric tons in Belgium, France, the Netherlands, Spain, Canada, the United States, Taiwan, and Brazil, Taiwan was (remarkably) the only one of the 27 countries listed in which imports exceeded domestic production. The United Kingdom was the only other country in which imports accounted for more than half of total supplies.

Only three countries, Norway, Mexico, and Taiwan, had no significant fresh exports, while nine countries reported zero or minimal imports. These included Greece, Hungary, Yugoslavia, Japan, Turkey, Argentina, Australia, Chile, and South Africa. Major apple producing countries use various rationalizations to block entry of competing apples while seeking free access for their exports. Clearly, such a position is not sustainable in the long run.

The United States, Belgium-Luxembourg, France, and the Netherlands were among the top ten importers and exporters. However,

TABLE 1.5. Analysis of Major Sources of Supply and Major Uses of Apples by Country and Region, 1989-90

	Production	Imports	Supply	Fresh Consumption	Fresh Exports	Processing	Withdrawal
Belgium-Lux	296,965	162,652	459,617	230,671	162,038	59,408	7,500
Denmark	92,500	49,622	142,122	102,973	4,149	35,000	0
France	1,851,550	113,900	1,965,450	1,033,350	665,100	200,000	67,000
Germany, W.	1,762,047	628,871	2,390,918	1,831,735	51,925	492,858	14,400
Greece	280,135	25	280,160	194,072	4,888	1,000	80,200
Italy	2,081,000	73,150	2,154,150	1,299,750	253,100	522,500	78,800
Netherlands	375,000	198,500	573,500	295,761	202,500	68,310	6,929
Spain	685,700	126,550	812,250	675,700	11,550	125,000	0
U.K.	342,550	474,459	817,009	750,545	29,012	16,630	20,822
TOTAL EC	7,767,447	1,827,729	9,595,176	6,414,557	1,384,262	1,520,706	275,651
Austria	254,700	14,150	268,850	266,350	500	2,000	0
Norway	59,683	42,197	101,880	80,029	0	9,921	11,930
Sweden	85,400	80,778	166,178	157,773	1,405	7,000	0
Switzerland	255,580	4,828	260,408	130,677	201	129,530	0
W. Europe	655,363	141,953	797,316	634,829	2,106	148,451	11,930
Hungary	954,500	0	954,500	295,500	326,500	332,500	0
Yugoslavia	523,000	0	523,000	293,000	30,000	200,000	0
E. Europe	1,477,500	0	1,477,500	588,500	356,500	532,500	0

Canada	500,000	105,000	605,000	323,000	87,500	194,500	0
Mexico	483,530	11,000	494,530	258,258	0	236,272	0
U.S.	4,398,000	117,250	4,515,250	2,229,000	340,000	1,946,250	0
N. America	5,381,530	233,250	5,614,780	2,810,258	427,500	2,377,022	0
Japan	1,057,000	0	1,057,000	842,500	1,400	213,100	0
Taiwan	15,380	104,039	119,419	118,668	0	750	0
Turkey	1,850,000	0	1,850,000	1,673,459	84,042	92,500	0
Asia	2,922,380	104,039	3,026,419	2,634,627	85,442	306,350	0
Argentina	1,067,500	0	1,067,500	222,500	225,000	620,000	0
Australia	332,500	0	332,500	193,525	24,475	114,500	0
Brazil	360,000	100,000	460,000	456,950	3,050	0	0
Chile	741,000	0	710,000	110,000	327,500	272,500	0
New Zealand	409,938	3,120	413,058	56,058	210,207	146,293	0
S. Africa	558,600	0	558,600	204,482	221,618	132,500	0
S. Hemisphere	3,438,538	103,120	3,541,658	1,244,015	1,011,850	1,285,793	0
Total Reported	21,642,758	2,410,091	24,052,849	14,326,786	3,267,660	6,170,822	287,581

SOURCE: USDA, FAS. Fresh Deciduous Fruit Reports. Washington, D.C. (unpublished, occasional attaché reports).

there is no way to know how much of the Belgian or Dutch trade is in transit between third countries. The United States had by far the largest allocation of its supply to processing, almost two million metric tons. Next, in order, were Argentina, Italy, West Germany, and Hungary, which all figure prominently in the world apple juice business. Withdrawals were highest in Greece, Italy, and France.

Western European countries were by far most dependent on imports as a source of supply. The United Kingdom imported almost 60 percent of its apple supplies, Norway and Sweden over 40 percent, and Belgium-Luxembourg, Denmark, and the Netherlands over 30 percent (Table 1.6). Taiwan was exceptional in importing over 87 percent of its supplies due to its limited domestic production.

Chile, New Zealand and South Africa exported between 40 and 50 percent of their supplies, almost all of which were domestically produced. Belgium-Luxembourg, France, the Netherlands and Hungary all exported more than one third of total supplies. The United States, while a major trader by volume, imported only 2.6 percent of its supplies and exported only 7.5 percent. However, imports of processed apple products are not included.

In general, the regions with the lowest percentage of supplies consumed fresh and the highest percentage processed were in North and South America. In Europe, only Switzerland, Hungary, and Yugoslavia showed a similar allocation. However, the percentage consumed fresh domestically tended to be lowest in countries like Chile, New Zealand, Argentina, South Africa, and Hungary where the combined proportion of supplies exported or processed was large.

The fresh consumption data used in the USDA balance sheets can be divided by population in each country to derive an estimate of fresh disappearance per capita (Table 1.7). Disappearance includes product consumed, plus shrinkage, waste and other nonconsumptive use. Nonconsumptive losses may be very high, 30 to 40 percent, in countries with poor handling, storage or distribution technology.

For the selected countries, fresh apple disappearance averaged 26.6 pounds per capita in 1989-90. The heaviest per capita use was in the three neighboring states of central Europe, Austria, West Germany, and Hungary, and in Turkey. Per capita disappearance exceeded 40 pounds in most other European countries. The exceptions were Spain, the United Kingdom and Yugoslavia. In the Western and Southern hemispheres, no country had disappearance above 40 pounds per capita. Only the former British colonies, Canada, Australia, and New Zealand, exceeded 20 pounds per capita. While per capita disappearance of fresh apples in the United State reversed its long-term decline in the mid-1970s, it was still less than 20 pounds in 1989-90.

Clearly, availability of local supplies and established consumption patterns contribute to increased fresh disappearance of apples. In contrast, export opportunities in higher income markets (e.g., for Chile in Western Europe) lower domestic disappearance. Disappearance tends to be lower in tropical or semitropical climates. Larger supplies, lower prices and rising incomes in the 1990s may tend to push world per capita disappearance upwards in the next decade.

TABLE 1.6. Share of Supply and of Major Uses of Apples by Country and Region, 1989-90

	Production	Imports	Supply	Fresh Consumption	Fresh Exports	Processing	Withdrawal
Belgium-Lux	64.6	35.4	100.0	50.2	35.3	12.9	1.6
Denmark	65.1	34.9	100.0	72.5	2.9	24.6	0.0
France	94.2	5.8	100.0	52.6	33.8	10.2	3.4
Germany, W.	73.7	26.3	100.0	76.6	2.2	20.6	0.6
Greece	99.99	.01	100.0	69.3	1.7	0.4	28.6
Italy	96.6	3.4	100.0	60.3	11.7	24.3	3.7
Netherlands	65.4	34.6	100.0	51.6	35.3	11.9	1.2
Spain	84.4	15.6	100.0	83.2	1.4	15.4	0.0
U.K.	41.9	58.1	100.0	91.9	3.6	2.0	2.5
TOTAL EC	81.0	19.0	100.0	66.9	14.4	15.8	2.9
Austria	94.7	5.3	100.0	99.1	0.2	0.7	0.0
Norway	58.6	41.4	100.0	78.6	0.0	9.7	11.7
Sweden	51.4	48.6	100.0	94.9	0.8	4.2	0.0
Switzerland	98.1	1.9	100.0	50.2	0.1	49.7	0.0
Total W. Europe	82.2	17.8	100.0	79.6	0.3	18.6	1.5
Hungary	100.0	0.0	100.0	31.0	34.2	34.8	0.0
Yugoslavia	100.0	0.0	100.0	56.0	5.7	38.2	0.0
Total E. Europe	100.0	0.0	100.0	39.8	24.1	36.0	0.0

Canada	82.6	17.4	100.0	53.4	14.5	32.1	0.0
Mexico	97.8	2.2	100.0	52.2	0.0	47.8	0.0
U.S.	97.4	2.6	100.0	49.4	7.5	43.1	0.0
Total N. America	95.8	4.2	100.0	50.1	7.6	42.3	0.0
Japan	100.0	0.0	100.0	79.7	0.1	20.2	0.0
Taiwan	12.9	87.1	100.0	99.4	0.0	0.6	0.0
Turkey	100.0	0.0	100.0	90.5	4.5	5.0	0.0
Total Asia	96.6	3.4	100.0	87.1	2.8	10.1	0.0
Argentina	100.0	0.0	100.0	20.8	21.1	58.1	0.0
Australia	100.0	0.0	100.0	58.2	7.4	34.4	0.0
Brazil	78.3	21.7	100.0	99.3	0.7	0.0	0.0
Chile	100.0	0.0	100.0	15.5	46.1	38.4	0.0
New Zealand	99.2	0.8	100.0	13.7	50.9	35.4	0.0
S. Africa	100.0	0.0	100.0	36.6	39.7	23.7	0.0
Total S. Hemisphere	97.1	2.9	100.0	35.1	28.6	36.3	0.0

SOURCE: TABLE 1.5

TABLE 1.7. Population and Per Capita Disappearance of Fresh Apples by Country and Region, 1989-90

	Population (1990) (millions)	Per Capita Disappearance (lbs)
Belgium-Luxembourg	10.2	49.7
Denmark	5.1	44.5
France	56.3	40.4
Germany, W.	60.5	66.8
Greece	10.1	42.6
Italy	57.2	50.1
Netherlands	14.8	44.0
Spain	39.1	38.0
U.K.	56.9	29.1
Total EC	310.3	45.6
Austria	7.5	78.0
Norway	4.2	41.9
Sweden	8.4	41.6
Switzerland	6.5	44.3
Other W. Europe	26.6	52.6
Hungary	10.5	61.8
Yugoslavia	23.8	27.1
E. Europe	34.3	37.7
Canada	26.6	26.7
Mexico	87.3	6.5
United States	249.8	19.7
N. America	363.7	17.0
Japan	123.3	15.1
Taiwan	20.6	12.7
Turkey	56.0	65.9
Asia	199.9	29.0

	Population (1990) (millions)	Per Capita Disappearance (lbs)
Argentina	32.3	15.2
Australia	17.0	25.0
Brazil	150.2	6.7
Chile	13.2	18.4
New Zealand	3.4	37.0
S. Africa	35.7	12.6
S. Hemisphere	251.8	10.9
TOTAL REPORTED	1,186.7	26.6

SOURCE: Population: Bulatao, Rodolfo A., Edward Bos, Patience W. Stephens and My T. Vu. *World Population Projections,* 1989-90 edition. World Bank, Johns Hopkins University Press, Baltimore, MD, 1990.
Disappearance: TABLE 1.5, Fresh Consumption.

Chapter 2

Production Systems and Costs

The life cycle of the apple plant in nature involves the sequence of bud formation, blossom, pollination, and the development of a sac containing the seeds for the next generation of plants. In due course that sac would reach maturity, be eaten by birds or other animals, or fall to the ground. As the sac disintegrated, the seeds inside would be scattered and those finding a favorable site would germinate, become plants and begin the cycle again.

We can assume that at some time in prehistory, humans discovered that the sac was good to eat. More recently, humans learned how to domesticate apple plants so the sacs could be picked more conveniently. Since cultivation began, the object has been to maximize the amount of fruit flesh produced by the tree with little concern for continuation of the life cycle of any particular plant.

Over the centuries, humans have learned how to intervene in nature to better meet their desire for plentiful fruit. Apples were cultivated in gardens and orchards in pre-Confucian China, and in the ancient Greek, Roman, and Persian empires. Irrigation, fertilization, pruning, thinning, tree training and other arts were commonly practiced.

With the advent of the printing press in the mid-fifteenth century and the focus on scientific knowledge as a result of the Industrial Revolution, scientists began to study and record their plant experiments. As the empires of Britain, Spain, France, Holland, and Portugal fanned out across the globe, plants were brought back and propagated in artificially generated environments. Much of the expertise gained was applied to improvement of European agriculture. The first horticultural societies were formed to exchange such information.

In the early nineteenth century, it was considered inappropriate for universities such as Oxford or Cambridge in England, or Yale or Harvard in the United States to pursue the study of plant or animal husbandry. While these studies gradually became more respectable, only a few isolated scientists paid attention to advancing horticultural science. In the United States, however, the Morrill Act of 1862 set up the land grant university system with the express purpose of training the working classes in the agricultural and mechanical arts. The Hatch Act of 1887 authorized federal grants to states for research while the Smith-Lever Act of 1914 set up a joint federal-state extension service. Thus, over half a century, this unique network of teaching, research and extension was set in place.

Because of this system, U.S. agriculture was able to rapidly adapt to the technological breakthroughs in other sectors. It adapted labor-saving machinery and equipment, tapped into new sources of power and energy, widely adapted chemicals for fertilizer, herbicides and insecticides, and steadily increased yields through scientific breeding. Other advanced countries developed alternative systems for applying science to agriculture. In the 1970s and 1980s, as the speed and ease of international communications increased dramatically, breakthroughs in one country could be rapidly detected and transferred to others. It seemed that there was no limit to the ways in which science could be used to intervene in nature for the benefit of humans.

However, in the 1990s, the use of science to intervene in nature for the benefit of humans is increasingly being questioned. The concept of making the desert bloom through irrigation is running into increasing conflicts with other users of water, such as energy generation, water transportation and migratory fish. The chemicals which appeared to be the solution to insects, pests and diseases in fruit have had problems with buildup of resistance, hazards for workers and unhealthy residues in food.

Much of the current technology used in apple production was derived in an era when the ends of increased production, improved appearance and improved quality were automatically assumed to justify the means required to achieve them. A number of current practices are being challenged within the industry by growers and scientists and outside the industry by environmentalists, consumer advocates and others. At the same time that economic pressures are encouraging growers to adapt new technologies in production, external constraints are setting limits on the methods that can be used.

THE ECONOMICS OF APPLE PRODUCTION: UNDERLYING PRINCIPLES

The economics of apple production, as of any perennial, can be extremely complicated. Commercial growers range from part-time farmers with limited acreage, to full-time family farmers, to multinational multiproduct corporations. It is not feasible to deal with all possible economic conditions that may face producers in such different circumstances. However, some general principles may be of value both to those entering the business for the first time and to those wishing to review the viability of their existing business.

The first principle of importance is that a considerable lag exists between the time that investment in an orchard planting begins and the time fruit production begins. While technologies have been developed to speed up the process of initial fruit bearing, it still requires at least a decade before an orchard will reach full bearing. Thus, the costs of establishment of an orchard must be charged to the orchard during its productive years. Even a grower who takes over an existing productive orchard will have paid the seller for the development costs already incurred by the developer, unless that developer is willing to take a capital loss.

The second principle is that some of the costs are incurred by the entire business, some by the apple enterprise, some by specific apple blocks and some by individual trees or apples. For example, a typical family farm may involve the owner in a very substantial capital investment, considerable managerial and supervisory duties, and personal labor of the owner, a spouse or other family members. Some way must be found to allocate these different costs to the apple enterprise. Most fruit farms grow more than one fruit. Apple growers may also grow pears, peaches, sweet cherries, grapes, or other fruits. Different apple blocks may have been planted with different varieties, different densities and different support or irrigation systems, and require different horticultural operations. Specific trees may be treated differentially and, in harvesting, specific apples may be picked at different times because of their level of ripeness. Somehow, a common basis must be found for accumulating these costs and comparing them with projected or actual revenues derived from each tree, block or orchard. Economists have developed a number of accounting conventions for solving this problem.

A third principle is that the actual output of a given tree, block or orchard will tend to vary widely from year to year due to the combined effects of winter freezes, spring frosts, pollination weather and the incidence of pests, diseases and other natural hazards. The smaller the unit, the greater the percentage variability. For example, the percentage variability of output from individual blocks will tend to be higher than that of the entire orchard. In turn, the percentage variability of output from a single orchard will tend to be greater than that for a region.

To compound the problem, apples, like many perennials, are subject to alternate bearing, the phenomenon where an above average crop tends to be followed by a below average crop and vice versa. Since much of the costs of apple production must be incurred before the yield for any given year is known, the average costs per unit of output will vary widely from year to year. To complicate the issue further, the factors which cause yield to be higher or lower in any one orchard will tend to have a similar effect on neighboring orchards. Thus, if a grower has an above average crop in any year, it is highly likely that the crop in the district or region will be above average. Given normal demand conditions, such an above average crop will lead to a below average district price. Thus, the same phenomenon which may reduce the average unit cost of production is likely, at the same time, to lower the average unit price. Conversely, in the same season that unit costs of production are higher, district price is also likely to be higher. The implication of this predictable variability in average unit cost, average unit price, average unit profit and total profit is that a prudent apple grower will expect to face losses on a regular basis because of uncontrollable factors. Thus, the same prudent grower will not seek to make a profit in any one year but rather over a reasonable span of years during which gains might be expected to offset losses.

A fourth principle of relevance is that apples are not a homogenous output. The quality and the market value vary widely based on size, grade, storability and other characteristics. Equally important as total output of an orchard is the percentage of desirable grades and sizes included in that output. For example, in the United States, one thousand tons of apples qualifying for fresh sale would, on average, generate more than three times the revenue of one thousand tons sold for juice (USDA, NASS). Even within the fresh category, the most desirable fruit may sell for three to four times the price of the smaller, less desirable fruit. To complicate the issue, increased costs in pruning, thinning and other practices may be required in order to increase the quality of the output. Thus, the grower may actually increase revenue and profits by increasing costs.

A final principle affecting the determination of profitability is that seasons have extensive overlap. For example, at the end of harvest, a grower must prepare the orchard for winter. The expenses incurred can be considered as the first outlays for the next season's crop which will be harvested twelve months later. Before that harvest, the grower must incur various costs of pruning, spraying, thinning, etc. Even though harvest may be completed in October, apples may be packed, stored and sold at any time during the next twelve months. Thus, the income flow from sales, and the expenses related to those sales, may continue for two years from the time operations on any crop began. In a typical month, say October, a grower may be winding up the books on one crop, completing harvesting of the next crop and beginning to incur expenses for an upcoming crop.

The above principles are not meant to intimidate the serious apple producer. However, they are a warning of the complexity of the economics of apple production, and of the very careful accounting needed to assess the viability of an orchard operation. An early example of this conceptual approach to controlling costs can be found in Stanton and Dominick (1964).

ESTABLISHMENT COSTS

To establish an apple orchard requires an initial investment in land, trees, an irrigation system, associated buildings, facilities, machinery and equipment. Within any one of those categories, the producer has wide latitude in how much to spend.

For example, the suitability and price of land for orcharding varies with the potential productivity of the site. Suitability is affected by soil, slope, wind exposure, air drainage and other natural factors. Price can also be affected by the alternative uses of the land. For example, the same type of site will be less expensive in a rural farm setting than in an urban or scenic location where the site might be used for housing development, tourism or other high-value uses. In addition, price can be affected by the amenities near the site. Land will tend to be more valuable where water, power, telephone and other utilities are readily available and where the distance to suppliers or packing houses is short.

A major investment decision involves the number of trees to plant per acre. While small quantity discounts are available, the investment in trees will be approximately twice as large when planting 600 trees per acre as for 300 trees. The cost of planting will also vary directly with the number of trees planted. Costs will also vary with the variety of tree, size of tree, rootstock, and other quality characteristics.

Irrigation systems come in many forms. The traditional rill (or gravity flow) irrigation has increasingly given way to fixed undertree or overtree irrigation systems. In areas where water is scarce or expensive, drip irrigation has become more popular. For small trees in dense plantings directed microsprinklers may be desirable. Sprinkler systems can also be equipped to handle chemical sprays. In larger plantings, the sprinkler system includes an automated control panel which permits irrigation of different blocks for preset periods on a specified cycle.

Some of the land within the orchard may not be useable for tree planting because of ravines, rocks, frost pockets, streams or other hazards. In addition, some of the land will be required for buildings, roads, windbreaks, or access space. Savings on these costs may be made if a new block or blocks are being developed next to an existing established orchard. Some developers attempt to economize on scarce capital by phasing in the construction of needed buildings over more than one year. Buildings are normally required for vehicle and machinery repair and storage. The durability and coverage of storage will vary with the severity of weather conditions faced, from simple lean-tos to climate-controlled buildings. In hot, desert climates, no covered storage may be required. In some parts of the United States, worker housing units may be required in order to attract needed labor.

Windbreaks have been common in areas exposed to strong winds because of damage to trees, fruit damage or fruit drop. However, windbreaks have become more critical in newer high-density plantings of dwarf or semi-dwarf trees which are more easily blown over. Natural windbreaks such as Lombardy poplar, Russian olive, and Chinese elm are most frequently used. However, they eventually compete with fruit trees for water and soil nutrients. Many newer plantings now employ windbreaks of wire mesh interlaced with plastic. Windbreaks are also of increasing economic importance in high density plantings which involve a large initial capital outlay and depend on rapid production to generate quick cash flows.

The major investment in vehicles is for a pickup and tractor; in equipment, a PTO blast sprayer. Minor investments may be required in a mower, weed sprayer, fertilizer spreader, trailer and forklift. Costs can be reduced by buying used or reconditioned items and by phased purchases as items are needed.

This multiplicity of orchard establishment decisions has a major direct effect on the future viability of the orchard. The initial capital outlay must be recovered from the future earnings of the orchard before the owner can begin to earn a positive return on investment. The establishment costs must be financed either from the owner's own resources or by borrowing.

Interest will have to paid on borrowed funds. On a $10,000 debt per acre at a flat 10 percent interest rate, annual interest charges alone will be $1,000 per acre. In many cases, interest charges can be well above 10 percent. Some lenders favor variable rate mortgages which fluctuate with the prime rate or some other index. These rates have, on occasion, exceeded 20 percent. Variable rate mortgages add to the riskiness of an orchard investment. In some countries, governments have provided subsidized credit for orchard development either in reduced interest rates or in subsequent tax breaks on interest expenses. Such subsidies reduce the riskiness of the orchard investment.

However, even though the owner uses his or her own capital to develop an orchard, and does not pay interest to a lender, there is a real cost of that capital. For example, a million dollars used to establish a 100 acre orchard in 1991 could have been invested at minimum risk in government bonds maturing in ten years at close to 10 percent annual rate of interest. Thus, the owner is forgoing a secure annual income of $100,000 before taxes. Even this comparison understates the opportunity cost to the owner. It would be more appropriate to compare possible returns from orcharding with those from alternative commercial businesses which, in general, will yield a higher return than government bonds.

The direct expense burden on the future orchard will vary with the establishment cost per acre and the actual or imputed interest rate. However, the level of each is critical. For example, an investment of $10,000 per acre at a fixed rate of 10 percent will incur just about half the interest expense of an investment of $14,000 per acre at a fixed rate of 14 percent ($100,000 versus $196,000). Failure to control these two costs can doom the entire enterprise.

However, the initial establishment decisions also have indirect effects on future costs and revenues that may be equally important. There is likely to be little correlation between the least cost establishment decisions and the long-run least cost to the orchard. For example, greater care in site selection, adequate soil testing and soil preparation before planting will cost more initially but will generate benefits for the life of the orchard.

The choice of rootstock has immense long-term consequences. Rootstocks vary in their vigor, early bearing, size control attributes, susceptibility to fire blight, mildew, phytophthora and other diseases, winter hardiness, and other important characteristics. Rootstocks derived from the research work at East Malling, Kent, England have become the industry standard, especially in semidwarf and dwarf plantings, but are being challenged by newer versions from the Cornell-Geneva apple rootstock breeding program, and from Poland, Russia and Sweden (Cummins and Aldwinckle, 1991).

The choice of rootstock, combined with the decisions on number and spacing of trees per acre, has a major impact on future costs and returns. The rapidity of early bearing, how soon peak production is reached, pruning requirements, access to the orchard and labor conditions are all affected. For example, a tree on M-26 (M indicating Malling origin) rootstock will reach about half the height and occupy two-thirds of the space of a tree from seedling rootstock. The smaller trees will come into bearing much earlier, will require more intensive managerial expertise, will require extensive supports, but will generally be easier to prune, thin, spray and harvest. However, while the smaller trees will generate a positive cash flow more rapidly, their peak production per acre is likely to be less than that from larger trees. In addition, a larger number of trees per acre on vigorous rootstocks can create serious control and access problems as the trees mature.

Much of the debate over the appropriate number and size of trees has been framed in terms of "high-density" plantings versus "traditional" plantings (Fisher, 1966). However, the definition of what is considered "high-density" has changed dramatically in the 1970s and 1980s and is likely to continue to change into the twenty-first century (Kortlave, 1991). It also varies by location. What is considered high-density in North America may be relatively low-density in parts of Europe. With their high land and operating costs, some European orchardists are experimenting with upwards of 4,000 trees per acre.

The individual orchardist must decide what is the best combination of tree numbers, size, and spacing for his or her goals and limitations. Two major pressures for increased tree numbers are the need for earlier cash flow to offset rising establishment costs and the need for growers to be able to respond more flexibly to changes in popularity of different varieties. In the past, commercial trees could remain in production for 50 or more years. In the 1990s, few new plantings contemplate an expected life of more than 20 years. In the early twenty-first century, the time horizon for a new orchard may be closer to 10 years.

Other establishment decisions need to be addressed with the same analytical detail as is required by the decision on planting density. The irrigation system must be tailored to the existing water supply, to the needs of the trees, both when newly planted and as they grow to maturity, and to the prevailing soil and climatic conditions. For example, a drip irrigation system with an emitter for each tree will ensure water to each tree in even the densest planting. However, unless the water supply is relatively pure, there will be need for constant monitoring to prevent sediment, for increased filtration, and for increased maintenance. An overtree sprinkler system is more easily adapted for climate control and spray applications than an undertree system, but it can cause trouble in areas or with varieties that are sensitive to fire blight.

Frost protection systems remain important in many growing areas. Smudge pots and other heating units are running into increasing environmental opposition because of the smoke, smell and hazards of reduced visibility. Overtree sprinklers can be used to seal the buds in a cocoon of ice which raises the temperature around buds by several degrees. However, large commercial growers in frost sensitive areas are forced increasingly to invest in wind machines which raise air temperature nearer the ground by replacing that air with warmer air above.

While most establishment costs need to be incurred before or during the first year of planting, some items may be delayed until fruit production actually requires them. Purchase of items such as storage sheds, worker facilities and harvesting equipment such as bins, ladders, picking bags, etc., may be phased in as needed. In addition, certain orchard operations such as irrigation, fertilization and rodent control that take place in the early years of orchard development are frequently included in establishment costs.

PRODUCTION COSTS

Production costs for a particular crop can be estimated by keeping a record of the cost of resources applied directly to the production and harvesting of that crop and by allocating to that crop an appropriate share of establishment costs and other overhead costs. For management purposes it is often desirable to be able to compare the per acre costs within an enterprise for blocks of fruit of different varieties or different other characteristics.

Many analyses of production costs deal with the three categories of growing costs, harvesting costs and marketing costs (for example Kelsey, 1990). This is appropriate for many small owner-operators near enough to large population centers to market their own limited volume. However, in most larger operations, the marketing phase is treated as a separate profit center (for example Hinman et al., 1992; Lee, 1986; LePage and Jackson, 1988). In many of the major apple trading nations, too, marketing is a specialized function carried out by larger private firms or cooperatives on behalf of a number of growers. Marketing is removed from orchard operations, both in nature and location. In this book, marketing systems and costs are dealt with in a separate chapter. However, it is recognized that in many orchard enterprises many of the marketing costs are incurred within the business.

Another important distinction is between variable costs and fixed costs. Fixed costs occur whether or not the orchard produces a crop in any season. Fixed costs include orchard depreciation, interest on investment, land taxes, tractors, machinery and buildings. Note that in the example cited from Washington (Dickrell et al., 1987), fixed costs amounted to over 40 percent of estimated total costs (Table 2.1).

Tractors and machinery costs classified as fixed costs include depreciation, interest on the average investment, and insurance. Fixed costs are allocated to the apple enterprise based on the average hours per acre each item is used in apple production. Variable costs of tractors and machinery include fuel, oil and repairs. The overhead item includes utilities and legal and accounting services related to orchard operations, while interest on operating capital assumes the orchardist borrows for current operations at a 10 percent interest rate.

Variable costs are separated for convenience into preharvest, harvest and postharvest costs, representing approximately 40, 12 and three percent of total costs. Postharvest costs occurring in the fall can be thought of either as clean up from the crop just harvested or as the necessary preliminary for the crop to be harvested the following year.

Under Washington's arid climate, the largest single expense is for labor (including pickers), just slightly over 30 percent of total costs. The second highest expense is for chemicals applied as fertilizer, insect sprays, thinning agents or growth regulators, over eight percent of total costs. Of course, much of the use of tractors and preharvest labor, and all the use of sprayers is involved in application of chemicals. Thus, the fertilization and spray programs overall

TABLE 2.1. Estimated Cost of Producing Apples in the Wenatchee Area of Washington State, 1987, $ per acre

Major Cost Categories	Item Description	$ per acre	Percent[1]
Variable costs, Preharvest			
	Fertilizer, chemicals, and spray materials	268.43	6.53
	Labor	857.34	20.87
	Irrigation electricity	40.00	0.97
	Irrigation charge, water	45.00	1.10
	Irrigation repairs	15.00	0.37
	Tractors	16.61	0.40
	Machinery	221.32	5.39
	Overhead	107.57	2.62
	Interest on Operating Capital	91.90	2.24
	Subtotal, Preharvest	1,663.17	40.48
Variable costs, Harvest			
	Pickers	330.00	8.03
	Labor, other	39.60	0.96
	Hauling	90.00	2.19
	Tractors	12.61	0.31
	Machinery	9.25	0.23
	Subtotal, harvest	481.46	11.72

Variable costs, Postharvest		
Fertilizer, chemicals, and spray materials	68.55	1.67
Labor	33.00	0.80
Tractors	9.46	0.23
Machinery	3.42	0.08
Subtotal, postharvest	114.43	2.79
Total Variable Costs	2,259.06	54.98
Fixed Costs		
Orchard depreciation	150.00	3.65
Interest on investment	1,050.00	25.55
Land taxes	60.00	1.46
Tractors	140.39	3.42
Machinery	337.97	8.23
Buildings	111.23	2.71
Total Fixed Costs	1,849.59	45.02
Total Costs	4,108.65	100.00

SOURCE: Dickrell, Peter A., Herbert R. Hinman, and Paul J. Tvrgyak. 1987. *1987 Estimated Cost of Producing Apples in the Wenatchee Area*. Pullman, WA: Washington State University Cooperative Extension Bulletin 1472.

[1]Percentages may not add exactly due to rounding.

may account for twice that share of total costs. In more humid growing areas, the spray program will account for an even greater share, in some cases equalling the percent spent on labor.

MANAGING COSTS

While the cost of apple production has been presented above in terms of costs per acre, the object of these expenditures is to increase the value of the fruit produced more than the costs of production so as to increase the gap between revenues and costs. Thus, for each individual orchard operation, the grower must have at least an intuitive estimate of whether the cost of that operation will be less than the reduced losses or added revenues from the operation. For example, in a typical orchard operation, more than one spray may be required to control a particular insect, say codling moth. After one spray, 80 percent of the fruit may be protected from worms. After a second spray, 99 percent of the fruit may be protected. Whether a third spray would be warranted to save the remaining one percent of the fruit will depend on the cost of the spray versus the potential return from the fruit saved. Since the spray operations take place months before the average price of fruit is known, the grower must make an informed guess at what that potential return may be.

Not all orchard operations can be as easily characterized as the above example. Each additional spray not only increases the kill of the target insect, but it also increases the chance of buildup of resistance to that chemical, the possibility of hazards to workers, fruit, soil or water, and the possibility of increased loss of beneficial insects. In the case of chemical thinning designed to increase the size of fruit, the grower must weigh the loss of fruit numbers against the potential gain in average fruit size and the projected price for each size group. Despite the difficulties, growers must always be attempting to weigh marginal costs against marginal benefits.

Clearly, the path to better decision making must be based on better information. For example, the better a grower understands the life cycle and behavior patterns of the codling moth and monitors the growth of the codling moth population in the orchard, the more effectively fruit can be protected at minimum cost. The better the grower can understand and predict the market effects of each operation, the more likely that operation is to contribute to increased profits.

Many growers rely on benchmark studies of production systems and costs in their region as a standard against which to compare their production costs. These studies are conducted by local universities, extension agents, government officials or, occasionally, industry groups. They range from synthesized cost estimates for a representative farm (as shown in Table 2.1), to detailed case studies of selected types of orchards, to sample surveys of all types of orchards in a district. The representative farm approach is the least costly in time and money, the sample survey the most costly and difficult. However, the applicability of results to any individual orchard is likely to be least for the representative farm approach and most for the subcategories included in a comprehensive survey. In any case, the details on operations conducted, materials and labor used, and methods of accounting can be a valuable aid to every grower in creating an individual orchard accounting system.

Once growers decide that an operation must be conducted, they can best manage costs by increasing the efficiency of that operation. Adequate training, supervision and scheduling of labor is critical. Permanent labor must be effectively employed despite the vagaries of weather, irregular incidence of activities and differing levels of expertise. Temporary labor, whether paid by piece rate or per hour, may well require greater supervision. The effect of orchard operations tend to be cumulative. An inadequate pruning job may affect the future productivity of a tree. A hasty picking job may cause bruising which shows up later in the marketing channel and costs the grower revenue. Short-run saving on labor can be costly in the long-run.

Efficiency and effectiveness are equally important in the spray and harvest programs. Stanton and Dominick (1964) reviewed the spray programs of New York apple growers in 1962 to answer two questions – why the spray bill was as high as it was and what caused the big differences from farm to farm. They noted eight major factors that influenced control of costs.

1. *Level of disease and insect control desired.* They advocated setting controls at a level sufficient to prevent serious outbreaks, not attempting complete control.
2. *Level of control achieved.* Preventive programs make more sense than curative ones. Once disease or insects get ahead of a grower, only high cost materials can regain control.
3. *Size and quality of crop.* Growers tend to use a minimum spray program when crops are small but are more generous with heavy-bearing premium varieties. However, heavier spraying was often accompanied by average yields.
4. *Number, size, spacing and pruning of trees.* Air-blast sprayers are most effective with regularly spaced, smaller trees of uniform size, properly pruned.
5. *Type of materials used.* Dusts are more expensive than equivalent sprays. Preventive materials cost less than eradicants. However, the efficacy of the material used is most critical.
6. *Timing of sprays.* Cover needs to be provided when it is most needed and when it will be most effective.
7. *Method of application.* The more highly concentrated the spray, the more critical that equipment be working properly, nozzles be adjusted and coverage of trees be checked.
8. *Management skill of the operator.* The experience, knowledge and judgement of the decision maker is critical in balancing cost against level of control.

To these cost issues raised by Stanton and Dominick, the modern orchardist must add concerns about worker safety, the buildup of resistance from greater frequency of use of chemical sprays, soil and water pollution, aerial drift, risks to neighboring populations and restrictions on use of available sprays. As more and more national and regional laws are passed to regulate these external effects of orchard practices, orchardists have looked at alternative spray regimes such as integrated pest management (IPM), low input sustainable agriculture (LISA), or use of organic materials only. These alternative approaches will be discussed in a subsequent chapter.

Optimal harvest practices involve picking the highest value of crop possible from what nature has provided, at the least possible cost, and within the constraints imposed by length of harvest period, weather and the available work force. The length of period during which marketable crop can be harvested varies by variety. The harvest period for summer varieties such as Gravenstein tends to precede that of early fall varieties such as Golden Delicious and Red Delicious and later fall varieties such as Granny Smith and Fuji. Hotter weather tends to speed up maturity, while frost can cut off harvest prematurely. Nights that are not sufficiently cool delay full red color in Red Delicious, while cool weather may prolong the period during which fruit ripens.

In general, apples for long-term storage will be picked first, before the fruit has reached its climacteric. Apples for intermediate-term storage will be picked next and apples for immediate fresh sale last, at the full ripe stage. Each grower must determine which target market will be most profitable and decide how many passes through the orchard pickers must make to get all fruit of comparable maturity. If all apples on a tree or in a block ripened on the same schedule, one pass through the orchard would be optimal. The more selectivity the picker must exercise, the slower the picking process and the higher the unit cost. Where growers are rewarded for being more selective or penalized for not being selective, they will be more likely to make a number of passes through the orchard.

The greater the number of passes and the greater the selectivity required, the more pickers will be needed and the better their training and supervision will have to be. As in the spray program, management skill will be a key factor in balancing quality control and cost. If the orchardist runs into a constraint of picker numbers or time, a less than optimal compromise may be necessary.

The logistics of efficiently moving the picked fruit from the pickers' bags or baskets down the rows to bins, to collection points within the orchard, and out of the orchard to cool storage also requires considerable management skill. The movement must take place in such a way as to minimize any loss of fruit quality. From the moment the pristine fruit is taken from the tree, each handling, contact or transfer has the potential to reduce fruit quality. Unless all orchard workers are aware of the market consequences of careful handling, any savings in harvest labor can be subsequently offset by penalties for bruised or damaged fruit.

Chapter 3

Apple Consumption Patterns

The ultimate goal of the commercial apple industry must be to satisfy the varying needs of consumers. Unfortunately, too often that does not occur. In many cases, the apple industry has not used the existing expertise in government, universities or the private sector in identifying apple consumption patterns. In addition, because of the long lag between planning of an orchard planting and first commercial harvest, and the traditional reluctance of growers to remove trees once commercial production begins, very often mature orchards have not adapted to changes in consumption patterns. In general, the apple industry has regarded each year's crop as a gift of nature to be disposed of as profitably as possible.

However, the worldwide experimentation with newer varieties, the push for early bearing in high density orchards, and the shorter expected life of orchards means that producers can more precisely target production towards selected consumers. Thus, a better understanding of apple consumption patterns is absolutely essential to the success of modern orcharding.

PRINCIPLES OF CONSUMPTION

The principles governing apple consumption are the same as those governing consumption of all other foods. Food is consumed first to meet the body's minimal physiological needs, that is, for human survival. At a second level, food is consumed above this threshold to meet the energy and nutritional requirements of normal work and play. At a third level, food is consumed for the enjoyment, satisfaction or pleasure it gives. Populations in a subsistence mode will spend their scarce resources on grains such as wheat, rice or corn. It is only as they reach the second level that fruit will be added to the diet. In many advanced countries, where all kinds of food are plentiful and overeating is a problem, fruits of many species become a major part of the pleasure derived from food.

A few basic demand concepts which have been empirically tested in countless technical economic studies around the world can be very useful in interpreting quite sophisticated analyses of apple consumption (Tomek, 1972). Economists talk about the utility or satisfaction of a good. We need only note two properties of utility. First, utility will tend to decline for each additional unit of a good consumed. A consumer may enjoy their first apple of the day a lot, but will find the second or third much less enjoyable. At some point, the consumer will get no enjoyment from eating an additional apple. Second, because consumers crave choice, they will be less and less willing to give up other foods, for example, oranges, to get an additional apple. Conversely, the more apples a consumer can get, the more willing he or she will be to give them up for a substitute fruit.

The practical result of these properties for apples or any other good is the familiar downward sloping demand curve (Figure 3.1). The line DD indicates the number of pounds of apples a consumer will buy at each price level. When the price is 60 cents, the consumer will buy only 2 pounds, when the price is 20 cents, 4 pounds. Note that in this case, the consumer actually spends less on apples when price is low (80 cents) than when price is high ($1.20).

Apple growers (and others) sometimes mistakenly believe that the basic laws of utility and demand have been repealed when they compare returns between seasons. For example, an individual grower, a region, or, occasionally, the world market, may sell a larger volume of apples at a higher price in a particular year than in the previous year. Usually this occurs because in the higher price year, there is a shortage of substitute fruits. The demand curve will shift to the right, from DD to D_1D_1 in Figure 3.2. The consumer may buy 4.5 pounds of apples at 30 cents per pound, rather than 4 pounds at 20 cents. However, note that the new demand curve, D_1D_1, is still downward sloping to the right. The consumer still pays less for a greater quantity. The law of demand has not been

FIGURE 3.1. Typical Consumer's Demand for Apples

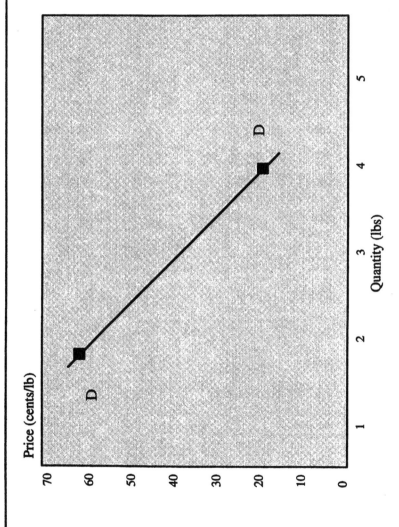

FIGURE 3.2. Effect of Shortage of Substitute Fruits on Consumer's Demand for Apples

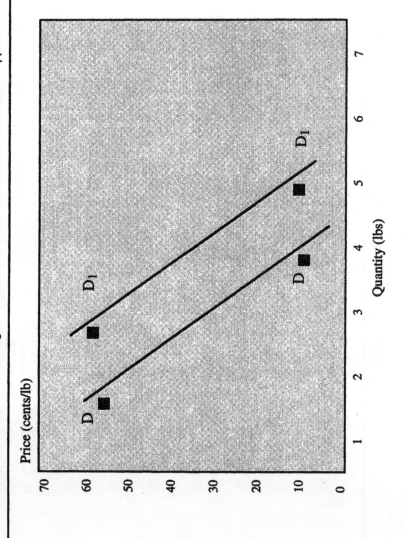

repealed. Demand can just as easily shift to the left if there is a glut of competing products, a quality problem, a health scare or some similar factor that dampens demand.

Three other factors can obscure the true nature of demand: population change, income change, and general inflation. For example, the U.S. population grows by about 2 million people annually. In a decade, the U.S. adds the equivalent of the entire population of Canada. In order to avoid distortions to the demand picture from population growth, consumption is usually analyzed on a per capita basis. Consumer income also affects demand. In general, the higher people's income, the more variety of fruits they will buy, the more particular they will be on quality, and the more additional services with their food they will require. Income can also be a proxy for changes in lifestyle over time. Accordingly, analysts usually attempt to separate out the effects of per capita income changes on consumption.

Finally, changes in general inflation (price levels) can be misleading. For example, the general price level in the United States as measured by the consumer price index in 1990 was about 50 percent above its level in 1980. Thus, if a consumer bought 4 pounds of apples at 20 cents per pound in 1980, we would expect that if demand had not changed, the same consumer would be willing to pay 30 cents per pound in 1990 for 4 pounds of apples. To examine the real price effect on apple consumption over time, analysts deflate (adjust) the reported prices by an index of general prices. Income is also similarly adjusted to remove the effect of general inflation.

A USEFUL DEMAND RELATIONSHIP

The demand for apples in any market can usually be explained in terms of factors that affect per capita consumption or price. For example, most of the variation in per capita consumption of apples can be explained by:

1. Apple price.
2. Price of competing fruits.
3. Income per capita.

If analysts wish to look at the effects of general inflation on apple demand, they can adjust prices and incomes by a suitable price index such as the Consumer Price Index published monthly by the Bureau of Labor Statistics.

Approximately the same formulation is useable whether one is concerned about the demand for all apples in a continent or the demand for a specific variety at a specific auction market in a particular city. The same formulation can be used to look at demand for different grades, pack-types, size groups or brands. There is much debate about the appropriate data to use for price, population, competing fruits, income and the price deflator in each case. However, the actual measurement is not as important to our purposes here as is awareness of what the key factors are which influence demand.

In measuring the strength of demand it is necessary to look at the joint effect of price and quantity. For example, many growers have responded to high prices of new varieties by heavy plantings. However, the ultimate test of the strength of demand is how large a quantity will be demanded at a specific price. For example, by my estimate, at $12 per box FOB, the Washington apple industry in 1990 could have sold 55 million boxes of Red Delicious, 15 million boxes of Golden Delicious and 6 million boxes of Granny Smith apples. Such a comparison gives a truer indication of the relative strength of demand than looking at price alone in any time period.

Determining the appropriate competing fruits is also difficult. For example, it is intuitively obvious that Washington Granny Smiths will compete with Granny Smiths from California and from the Southern Hemisphere. But what about other apples from Washington, or other green apples such as Newton, or other fresh varieties such as McIntosh or Empire? In all likelihood, these are not as direct substitutes for Washington Granny Smiths as other Granny Smiths. However, in some circumstances they may be. Only by accurate measurement can the relationship be determined.

While the theory of demand would lead us to expect that apple consumption would increase with income, there is reason to believe that the effect will become smaller as incomes reach a level where consumers have available the total quantity of food they can comfortably eat and a wide selection of foods among which to choose.

Given that scenario, one would expect little increase in apple consumption per capita in the richer countries such as the United States, Japan, Canada and the European Community as per capita income rises. In contrast, one would expect quite rapid increases in consumption as income increased in middle-income countries such as Taiwan, South Korea, Saudi Arabia and Brazil.

The basic demand relationship described above is a useful tool for analyzing demand. However, like all simple tools, it does not answer some critical underlying questions about why consumers and markets behave as they do. For example, Granny Smith apples are relatively green and tart. In contrast, Red Delicious are red and sweet. If we plot these color and sweetness characteristics on a two-dimensional diagram, we can see that these two apples fit into opposite and distinct quadrants (Figure 3.3). Red Delicious is clearly in Quadrant II, Granny Smith in Quadrant III. However, as seen in Table 1.3, about one quarter of world apple production was Red Delicious, less than 10 percent Granny Smiths. This would appear to indicate that many more consumers prefer a red, sweet apple than a tart, green apple. It does not indicate what the limits of the market for either type of apple might be.

Figure 3.3 also raises intriguing questions about where Gala, Braeburn, Fuji, Jonagold and other newer varieties fit into this schema. Clearly, if they are perceived to be quite distinct in color and sweetness from Red Delicious and Granny Smith, they can win an exclusive following. However, if they are identified too closely with an established variety, two things can happen. Either the total segment will increase, so consumption of the newer variety can expand without hurting the existing variety, or they can only grow at the expense of the competing variety. Given the proliferation of new varieties, some of both situations are likely to arise.

Economists have been attempting to explore how much people are willing to pay for the characteristics that are embedded in particular products. Using hedonic analysis, they are attempting to measure what is the value of different combinations of characteristics in a product. Some of these characteristics (such as color) may be observed directly. Others may be based on perceptions. For example, the presence of large bruises may lead the consumer to infer that an apple is insufficiently firm and to value it accordingly.

FIGURE 3.3. Positioning of Red Delicious and Granny Smith by Color and Sweetness Characteristics

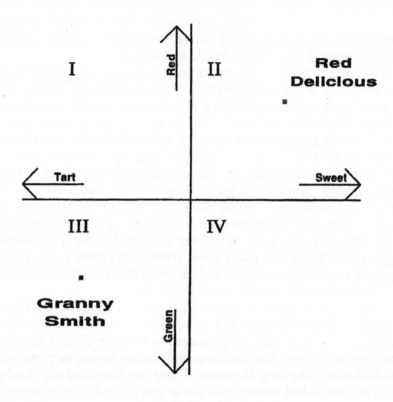

Hedonic analysis promises to help us understand which characteristics consumers value and how much they value them.

DEMAND AT DIFFERENT LEVELS

Up to this point, we have talked only about how demand for apples is affected at the point of final consumption. However, the producer is usually separated from the consumer in time, form and space. The producer and the marketing system can be thought of as adding time, form and space utility to the raw product. For example, a ton of apples may be harvested in September, stored for a month in a fresh pack warehouse, processed into bottled juice in October, stored for two months in the processor's warehouse, shipped to a retail store a week before Christmas, bought by a consumer on Christmas Eve and consumed at a family gathering on Christmas Day. Clearly the value of an apple is different in September than at Christmas, in raw form than as apple juice, at the orchard than at the consumer's home.

Many firms help link producers and consumers by adding time, form and space utilities. The activities of these firms will be discussed in subsequent chapters. Each activity both provides a service and generates a cost which must be paid for by the consumer. From the producer's perspective, the price received for the raw product is the price the consumer is willing to pay for the final product less the cost of the intermediate services. The gap between the consumer and producer price is referred to as the marketing margin.

It is possible to assess the demand for apples at the terminal wholesale, processor, fresh packer, grower or other level by assessing consumer demand and allowing for the influence of the marketing margin. Studies of demand at different levels of the marketing chain are said to be derived from final consumer demand. For many purposes, these derived demand studies may be valuable to growers, packers and processors. In general, data is weakest at the consumer level. Derived demand studies can be (and often are) conducted in cases where consumer data are not available.

Marketing margins themselves are of intense interest to growers and shippers, who are constantly concerned that the marketing system is taking an excessive share of the consumer's dollar for the services it performs. However, measurement of marketing margins is very difficult because of the long times and great distances over which apples now travel. The orchardist's share of the consumer dollar will tend to be higher the closer the orchard is to the final consumer and the more direct the marketing channel. For orchardists supplying a distant export market, the grower's share may be less than 20 percent of the consumer's dollar. Clearly, an understanding of the marketing system can be critical to the returns a grower receives. This is discussed more fully in later chapters.

WHAT WE KNOW ABOUT APPLE CONSUMPTION

As discussed in an earlier chapter, information on per capita consumption of apples worldwide is relatively sparse. Dividing world production by world population gives us some indication of the availability of apples per head of population (Table 1.1). Per capita availability averaged 5.6 kilograms in the 1961-65 period, slightly above post-World War II levels. It surged to 7.7 kilograms in 1969-71, and then rose only slowly to 7.8 kilograms in 1979-81 and 8.0 kilograms in 1986-88. Clearly, for the most recent two decades, per capita consumption of apples is likely to have grown little. Per capita consumption in Europe, by far the heaviest user of apples, also appears to have been relatively static. A similar situation prevailed in Japan during the 1970s and 1980s as the Japanese apple industry underwent major renewal (Heydon, 1981). In contrast, the per capita consumption in some of the newly industrialized countries of Asia and the Middle East, fuelled by imports, rose dramatically.

In the United States, per capita consumption of fresh apples showed a steady downward trend between 1940 and 1970 at the same time that per capita consumption of processed apples grew to meet demands for convenient, prepared foods (USDA, 1965b). A number of factors reversed these trends in the early 1970s. Under the Nixon price controls of 1971-73, fresh produce was exempt. This gave retailers an incentive to promote higher margin produce items in preference to processed foods. Produce items have become steadily more important to the profit picture of U.S. grocery retailers. Second, a series of reports on consumer health emphasized the value of increased consumption of fresh fruits and vegetables. Finally, the processing industry was battered in the 1970s by a sharp run-up in the cost of sugar and energy, by higher interest rates, and by increasing consumer concerns about the additives and nutritional losses associated with processed foods.

As a result, U.S. per capita consumption of fresh fruit, including apples, began a slow but steady recovery in the mid-1970s (Putnam, 1989), which continued into the 1990s. In contrast, the consumption of processed apples (excluding apple juice) stagnated during the same period. Per capita consumption of apple juice grew rapidly in that period from 6.36 pounds to 19.04 pounds raw product weight (Putnam and Allshouse, 1991). Canadian per capita consumption of fresh apples rose little in the 1980s, but apple juice consumption rose by almost one-third.

In the advanced countries, per capita consumption of fresh apples has not been very responsive to increases in the average income level. Tomek (1968) found that for the period, 1947-66, per capita consumption of fresh apples showed zero response to income growth. O'Rourke (1974) found that per capita consumption of Washington fresh apples declined slightly as income increased in the 1950s, but then became positive (although close to zero) in the 1960-70 period. In more recent work, O'Rourke (1990) found that the income effect on fresh consumption was still positive.

Data from occasional large cross-section studies of food consumption help explain why per capita consumption of fresh apples in the U.S. shows so little response to increases in average income. Most Americans have bought apples in the past year. About half report buying apples in the past week (USDA, SRS, 1966; Zind, 1987). The frequency of purchase is high in all locales, income groups and educational levels. However, the amount purchased is small.

For example, the USDA nationwide food consumption survey of 10,301 households in 1977-78 found that households on average used only 1.25 pounds of fresh apples per week (about 65 pounds per household per year). An analysis of this survey by income revealed that average use of fresh apples rose from 0.62 pounds among families with incomes of $0-$4,000 to 2.50 pounds among families with incomes between $49,000 and $74,000, before falling sharply for families with incomes over $74,000 (Figure 3.4). The frequency of use in any week rose from 32.3 percent among families with incomes of $0-$4000 to 55.7 percent among the $29,001-$49,000 income group. Frequency then fell to 53.8 percent in the $49,000-$74,000 income group and 42.4 percent among families with incomes over $74,000. Clearly, in the United States, frequency and amount of use peaked below the highest income levels as the more affluent purchased a wider selection of apple substitutes. This group is known to be among the most frequent users of novelty or

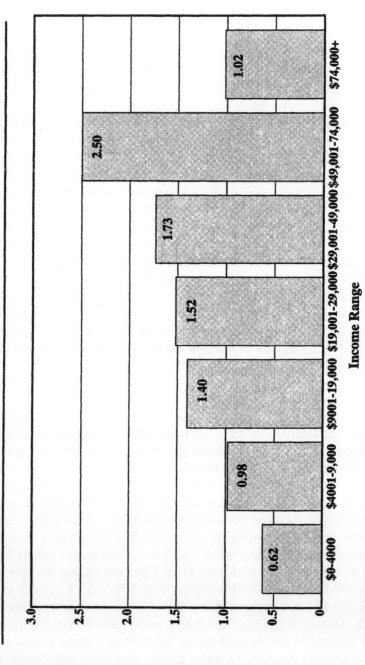

FIGURE 3.4. Average Weekly Use of Fresh Apples in the United States, by Income Class, 1977-78

Income Range

SOURCE: USDA, 1983. *Nationwide Food Consumption Survey, 1977-78.*

exotic fruits. However, they are also the most difficult group to survey effectively, so the data is subject to wider than normal error.

A second major factor affecting use of fresh apples is size of family. Frequency of use rose from 31.7 percent for one-person households to 57.3 percent for 4-person and 5-person households and 57.4 percent for 6-person households, before falling to 52.3 percent for larger households. The actual volume used per week rose from 0.51 to 2.41 pounds per household (Figure 3.5). Note, however, that the use per person fell steadily as family size increased. This may reflect the presence in larger households of more younger children, increasing income constraints and less wastage.

The U.S. Nationwide Food Consumption survey was analyzed by seven income categories and seven family size categories, giving 49 distinct market segments. However, the six market segments representing households with two to four persons and the middle-income groups, $9,001-$19,000 and $19,001-$29,000 accounted for 42 percent of the volume of apples used. The two middle-income segments with five persons accounted for a further 11 percent of total use and the two- to four-person households with $4,001-$9,000 income for a further 12 percent. Thus, almost two-thirds of all U.S. fresh apple use was accounted for by medium-sized, medium-income households.

The number of published studies which probe the behavior of apple consumers is fairly limited. In the 1960s and early 1970s, the U.S. Department of Agriculture had an active program to study consumer behavior with respect to fruit purchasing and consumption. With personnel changes and the emergence of other priorities, that effort was gradually phased out and has not been revived. In response to depressed apple prices after the 1969 record U.S. apple harvest, a number of universities cooperated in a Northeastern Regional Research Project, NE-88, "Future Adjustments in the Marketing of Selected Northeast Fruits and Vegetables." Several participated in surveys of fresh apple purchases (Frick and Toensmeyer, 1977; Cain and Shawaryn, 1976; Trotter and Brewer, 1977; MacGregor and Jack, 1979; Table 3.1).

These surveys showed that most fresh apples were bought in supermarkets, the dominant food outlet in the United States. The frequency of purchase showed a similar pattern across states, with

FIGURE 3.5. Average Weekly Use of Fresh Apples Per Person and Per Family in the United States, by Family Size, 1977-78

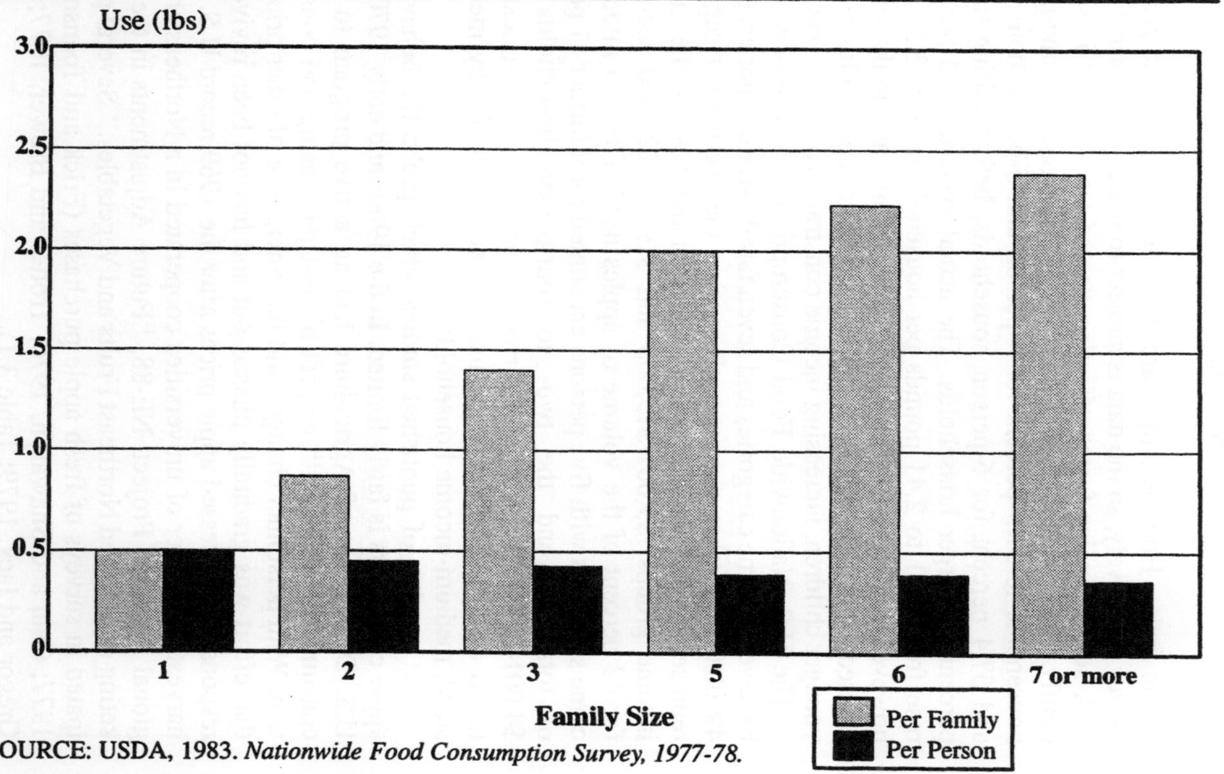

SOURCE: USDA, 1983. *Nationwide Food Consumption Survey, 1977-78.*

purchase every two weeks being most common. As one might expect, given production patterns, the favorite varieties were Red Delicious, Golden Delicious, and Winesap. However, McIntosh had a high preference rating in Pennsylvania. The four most reliable indicators of quality were: no bruises or blemishes, juicy and crisp, firmness, and color. The major negative factors were absence of these qualities and price. Consumers disliked apples that were mealy, soft or soggy, or that had external blemishes or internal blemishes such as browning. Consumers disappointed with fresh apples were most likely to substitute fresh oranges or bananas.

More recent consumer surveys indicate that apples remain the most frequently purchased fresh fruit grown in the United States. For example, a study by Vance Research Services for a major produce journal, *The Packer,* "Fresh Trends 1987," showed that 98 percent of respondents had purchased apples, and 97 percent of those purchases were within the past year. Only bananas, an imported fruit, had as high a response. Of those who purchased apples, seven percent did so twice a week or more, 36 percent weekly, 34 percent every two-three weeks, 14 percent monthly and 10 percent less than monthly. Only bananas were purchased more frequently. While these data confirm that about 40 percent of all households buy fresh apples at least weekly and 90 percent at least monthly, they do not indicate whether the more frequent users are more or less likely to try the newer varieties of apples. However, the Fresh Trends 1987 report indicates that frequent apple buyers purchase a higher proportion of Red Delicious and Golden Delicious, than of McIntosh or Granny Smith (Zind, 1987).

Data from the Household Budget Survey of the Republic of Ireland for 1987 present an interesting contrast to U.S. data (Central Statistics Office, 1989). Per capita income of the Republic of Ireland in 1987 was approximately 30 percent that of the United States. Sixty-seven percent of Irish households reported purchase of fresh eating apples and 18 percent of fresh cooking apples. This compared with 54 percent reporting purchases of fresh oranges, 53 percent bananas and 11 percent fresh grapes. As one might expect, the variety of fresh fruits purchased and the frequency of purchase was lower in a middle income country such as Ireland than in a higher income country such as the United States.

TABLE 3.1. Profiles of Fresh Apple Purchase and Use, Selected States, U.S., 1974-75

State	Delaware		Maryland	Pennsylvania		West Virginia
Time of Survey(s)	Fall 1974	Spring 1975	June 1974 and Jan. 1975	Oct. 1974	April 1975	Spring 1975
Frequency of purchase in supermarkets:						
Weekly	15.7	29.3	26.0	16.0	23.0	45.5
Twice weekly	2.3	1.6	3.1	0.0	3.0	6.8
Every two weeks	30.3	23.6	34.4	42.0	45.0	27.3
Monthly	32.6	21.1	22.6	30.0	24.0	9.1
Seasonally	19.1	24.4	13.9	12.0	5.0	11.4
Total	100.0	100.0	100.0	100.0	100.0	100.0
Favorite variety:						
Delicious, all	42.5	39.7	45.4	42.0	46.0	—
Red Delicious	—	—	24.5	37.0	38.0	37.0
Golden Delicious	—	—	20.9	5.0	8.0	46.0
Winesap	11.8	14.0	13.2[1]	11.0	14.0	23.0[1]
McIntosh	9.4	10.6	8.2[2]	32.0	29.0	4.0
Stayman	8.7	5.0	5.9	—	—	7.0[3]

Golden Grimes	1.6	5.0	4.5	—	—	12.0
Jonathan	—	—	3.1	6.0	4.0	3.0[4]
Other	4.0	0.6	2.1	9.0	7.0	2.0
None	22.0	25.1	17.6	—	—	20.0
Total	100.0	100.0	100.0	100.0	100.0	154.0[5]

Reliable indications of good quality:

Grower and place of origin	4.5	4.5	2.9	5.0	4.0	15.0
Shape and size of apple	10.2	8.1	6.8	9.0	7.0	26.0
No bruises or blemishes	21.9	25.7	25.1	24.0	23.0	63.0
Juicy and crisp	23.5	23.9	25.6	22.0	24.0	51.0
Color	9.9	9.3	8.6	10.0	11.0	34.0
Firmness	17.7	22.1	20.0	18.0	18.0	49.0
Variety name	6.8	4.8	6.5	8.0	9.0	25.0
Price	5.5	1.6	4.5	4.0	4.0	13.0

SOURCE: See Page 60, paragraph 3

[1]Including Rome Beauty [2]Including Greenings [3]Includes Winesap [4]Includes Northern Spy [5]Some respondents gave more than one answer.

The Household Budget Survey of Ireland analyzed expenditure on various items, including fresh fruits, by level of urbanization, region, income, household size, household composition, social status and family life cycle. Expenditure on fresh apples appeared to be most responsive to two items, changes in household size and income (Table 3.2). The largest households in the highest income quartile spent more than six times as much on fresh eating apples as the smallest households in the lowest income quartile. While household size appeared to increase expenditure on eating apples more than did income increases, it is notable that expenditure continued to rise even through the highest income group. This contrasts with the experience of the United States illustrated in Figure 3.4.

As one might expect, expenditure per person on fresh eating apples declined as household size increased. However, the decline was only 25 percent from one-person households to nine-person households. In addition, each additional child in a household was

TABLE 3.2. Average Weekly Household Expenditure on Fresh Eating Apples by Income Quartile and Household Size, 1987, £ Irish[1]

Income Group[2]/ Household Size	First Quartile	Second Quartile	Third Quartile	Fourth Quartile
1-2 Persons	.258	.324	.441	.490
3-4 Persons	.495	.619	.684	.817
5-6 Persons	.557	.766	1.004	1.295
7 + Persons	.982	.974	1.226	1.681

SOURCE: Central Statistics Office. 1989. *Household Budget Survey, 1987.* Volume I. "Detailed Results for All Households." Dublin, Ireland: Government Publications Sales Office.

[1]£ Irish, the currency of the Republic of Ireland, was selling at a small discount to the £ sterling in 1987.

[2]The first quartile had average weekly incomes of <£103.35, the second quartile £103.36 to 193.23, the third quartile £193.24 to £336.59 and the fourth quartile >£336.59.

associated with additional expenditure on eating apples equivalent to 83 percent of that associated with an additional adult. Rural farm households spent more on eating apples than either urban households or nonfarm rural households. Expenditure was highest among the professional and managerial social class, the employed and in family households with adolescent children.

Expenditure on cooking apples averaged about one-seventh of that on eating apples. It tended to rise with income and household size and to be higher in rural farm households and in households with adolescent children. Similar data are not available for the United States where eating and cooking apples are not generally distinguished. However, surveys consistently show most U.S. households reporting use of fresh apples for cooking, but not cooking them frequently.

Household consumption data on fresh apples are also available for British households, but direct comparison with the United States or Ireland is not usually possible. In "Household Food Consumption and Expenditure, 1989," the annual report of the United Kingdom National Food Survey Committee (MAFF, 1990), it was reported that the average consumption per person per week of all fresh fruit was 21.45 ounces at a cost of £0.567 sterling. The reported consumption of fresh apples was 7.26 ounces, fresh bananas 4.00 ounces and fresh oranges 2.99 ounces. Fresh pears (1.20 ounces) edged out fresh grapes (0.80 ounces) for fourth place. Fresh apples accounted for slightly more than one-third of all fresh fruit consumption. During the survey week, 48 percent of households reported purchasing fresh apples, 41 percent fresh bananas and 22 percent fresh oranges. Consumption of apples and bananas was relatively uniform throughout the year whereas consumption of oranges was much higher in the first half of the calendar year.

DEMAND ELASTICITIES

By their nature, cross-section studies such as those just described are designed to match the social, political and regional idiosyncrasies of each country. Accordingly, as we have seen, comparisons across countries are difficult. Demand elasticities are summary measures of the effect on consumption of changes in income or prices. Such elasticities can be used to make comparisons between different countries, different commodities and different periods.

Tomek (1968) estimated a direct price elasticity of −0.81 for fresh apples at the farm level for the crop years 1947-66. That means that each one percent increase in the real price of fresh apples would be associated with an 0.81 percent decrease in fresh apple use per capita, and by an overall increase in the total revenue to apple growers. Tomek found a cross-price elasticity with respect to fresh oranges of 0.10, that is, each one percent increase in the real price of fresh oranges was associated with one tenth of a percent increase in per capita use of fresh apples. Tomek found an income elasticity for that period of zero, that is, increases in real income had no measurable effect on per capita consumption of fresh apples. In an econometric model of the U.S. apple market for the period 1952-81, Baumes and Conway (1985) estimated a farm level elasticity for fresh apples of −1.135 and an income elasticity of 0.757. Their results suggest that since the 1960s, demand for fresh apples in the United States has become more sensitive to changes in both own price and income. An updated version of the Baumes and Conway model by O'Rourke for the 1969-84 period confirmed these trends.

The British Household Food Consumption and Expenditure survey previously discussed (MAFF, 1990) estimated retail demand elasticities from cross-section data on British households in 1989. It reported an own price elasticity for fresh apples of −0.29, and an income elasticity of 0.32. It reported cross-elasticities of −0.07 with fresh oranges and 0.01 with fresh pears, relatively small influences in both cases.

Heim and O'Rourke (1988) analyzed demand for fresh apples in Japan for the period 1963-86. They found a price elasticity of −0.57 and an income elasticity of −0.01. Heydon and O'Rourke (1982) analyzed demand at the import level for fresh apples in a number of Asian countries for the years 1960-78. They found price elasticities for Singapore, Hong Kong, Taiwan and China were all less than unity. Price elasticity for Malaysia was −1.17. Price elasticity for South Korea was positive (an unacceptable result). Income elasticity for all six countries was positive, but only greater than one for Taiwan.

In general, available elasticities suggest that the demand for fresh apples is boosted by lower prices and higher incomes. However, price elasticities are frequently less than one at the farm level, suggesting that increases in supply will lead to reduced total revenue to apple producers.

Other studies have indicated that demand for fresh apples varies by season and by variety. For example, Ben-David and Tomek (1965) found for the 1960-63 period that demand for New York apples was more elastic in the harvest season, September-November, and in the CA season, March-May, than in the intervening regular storage season. Thus, it would increase grower revenues to allocate some of the mid-season fruit to either early or late markets. Price (1973) looked at intraseasonal demand for Washington Red Delicious and Golden Delicious for the years 1957-68 and concluded that "with Red Delicious, late Spring sales should be increased while little is to be gained from changing the present Golden Delicious marketing pattern." Price (1973), Scott (1971) and O'Rourke (1974), in separate studies, found that for varying periods prior to 1971, the demand for Washington Golden Delicious apples was more price elastic than that for Red Delicious, meaning supply could be increased with less relative decrease in growers' revenue. However, using crop data for the 24 seasons 1964-87, O'Rourke (1990) found shipping point price flexibilities for Red Delicious, Golden Delicious and Granny Smith of -0.435, -0.246 and -0.19 and income flexibilities of 0.125 and 0.370 for Reds and Goldens. Clearly, the more widespread the use of a variety becomes, the more sensitive price is to additional supplies and the less it is helped by an increase in income.

Chapter 4

Warehousing and Packing: Technical Aspects

While the grower is initially responsible for the quality and quantity of apples available for marketing, it is at the next stage of the marketing system that major decisions are made which set the tone of the entire apple market. At this stage, apples are assembled, sorted by target market, and packed and stored to meet customer needs. Occasionally, a bulk merchandiser will buy apples "orchard run," that is, from the field without any additional preparation for market. In most growing areas, however, an extensive system has been developed to transform fresh apples from "orchard run" to "market ready."

The names of the firms and functions they perform vary. For example, grower-packers will have their own packing facility in or near the orchard where they pack their own fruit and that of neighboring orchards. The term grower-shipper usually indicates that the producer also arranges for the delivery of packed fruit to buyers. The terms packers, packinghouses, packing sheds, warehouses, handlers, etc., are used loosely to describe both the physical facilities involved and the types of firms. Some packers may also have their own sales desks. Others may employ a specialized sales or marketing firm.

SPECIALIZED WAREHOUSING AND PACKING FIRMS

To simplify discussion, we focus here on the specialized warehousing and packing firms which dominate the commercial fresh apple industry. These have plants located in the producing areas.

Each plant assembles fruit from many growers, presizes it at harvest, stores it until needed, sorts it, packs it and holds it ready for sale. They walk the difficult tightrope of competing in service to retain grower affiliates while competing in product quality and price for the goodwill of product buyers.

One of the most important decisions a grower can make is the choice of a warehousing-packing operation to handle his or her fruit. In many cases, growers have a preharvest contract to deliver all of their harvested fruit to a specific warehouse. That contract will usually stipulate how charges for the various functions performed will be assessed. In most cases, the grower retains ownership of the fruit. After the packed fruit and culls are sold, the grower receives whatever income is left after deduction of all warehouse charges. Occasionally, when shipping point price is less than all warehouse charges, the grower will get a bill for the difference. Thus, the warehouse has a fiduciary responsibility to look after growers' fruit. Trust between grower and warehouse is critical. Trust can be fostered if the grower understands how and why the warehouse operates as it does and if the warehouse makes its operations and activities open to growers.

While the most common form of organization among fresh fruit warehouses is likely to be the grower-packer, the biggest volume of fruit tends to be handled by firms which specialize in warehousing and packing. Many of these firms are private companies, but some are grower-owned cooperatives and others in certain countries are government-owned or controlled. Cooperatives and government-owned entities must meet the same economic and technical constraints as the private sector but may have additional responsibilities to cooperative members or taxpayers.

The number of multiplant firms has also been growing. Until the 1980s, most packers were specialized fruit companies seeking to gain a foothold in different parts of an apple-growing region or in neighboring states. More recently, a number of multinational, multiproduct firms such as Dole, Del Monte, United Brands (Chiquita), and Polly Peck have become factors in the apple warehousing and packing business. Single unit warehouses are concerned that the marketing muscle of these multinationals may give them an economic advantage in servicing growers.

OPERATIONS

Assembly is one of the most important operations conducted by fresh warehouses but also one that is most often taken for granted. From the moment that the picker disturbs the apple on the tree to the time it is resting in appropriate storage, the apple is exposed to numerous hazards. In general, the more rapidly it can be moved from tree to cooling, the fewer times it is handled and the more gentle the handling, the more the intrinsic qualities of the apple can be preserved. Assembly is complicated by the fact that it is squeezed into a few short weeks at harvest. The flow of fruit is often dictated by nature rather than by management. Many of the employees involved are temporary, hired to meet peak harvest needs. In the effort to move large volumes at high speed, concern for fruit quality may suffer.

In general, the larger the plant to be served, the further fruit must be hauled in order to meet plant capacity. The longer haul increases the risk of transit damage or the managerial attention needed to avoid damage. The actual costs of assembly will differ little whether the product is hauled by the warehouse or the grower. However, the grower has a more direct stake in maintaining quality than a warehouse employee.

While apples may be hauled to the warehouse in many containers and on many different vehicles, the most common commercial unit is a straddle carrier. Fruit is transferred from picking bags to wooden bins holding from 800 to 1000 pounds. These bins can be moved by forklift truck within the orchard. The straddle carrier is actually a trailer frame that can be driven over and hydraulically pick up 32 bins at a time. With a truck drive unit, these trailers can then be hauled by road to the warehouse at normal road speeds. Straddle carriers can haul 14 to 15 tons of fruit at one time.

The capability of rapidly assembling these large volumes of fruit from many growers at one time can easily lead to a bottleneck at the warehouse. Only a small percentage can be immediately packed because market demand for fruit from commercial warehouses at harvest time is limited by the plentiful supply of competing backyard, local and pick-your-own supplies in many areas. Rapid decisions must be made about each lot. Is there sufficient packable fruit to merit holding it or should it be consigned directly to the processor? Should it be run across the presizer before being stored? Should it be stored in regular or controlled atmosphere (CA) storage, in early or late CA storage?

These decisions can be less troublesome where horticultural experts employed by the warehouse are familiar with the orchard blocks from which the fruit came and have monitored those blocks during growing and harvesting. In some cases, the horticultural expert, in conjunction with the grower, may have agreed upon the timing and selectivity of picking of particular blocks to meet specific market needs. Virginia pioneered a rapid sampling system for bins arriving at the warehouse through use of strategically placed trapdoors. However, that system has not been widely adopted. In general, the earlier a warehouse can separate fruit of divergent quality, the less costs will be incurred subsequently in storage, sorting and packing.

To encourage growers to deliver only fruit of fresh-pack quality, many warehouses have instituted an in-charge per bin. Thus the grower must pay even if none of the fruit goes to the higher-value fresh market. This appears to have led to the direct sale of uniformly poor quality lots to processors, but to have done little to encourage growers to sort more carefully in the orchard.

Traditionally, apples not immediately packed were placed in storage still in their wooden field bins. They would gradually be removed from storage and packed to meet customer orders. However, warehouses found that to meet orders for a specified number of boxes of a particular size and grade of apple, they had to remove ten times as much fruit from storage to run across the packing line. The unsold fruit had to be held in packed box storage so that if quality problems occurred expensive repacking had to take place.

Presizing systems were introduced in the late 1960s to overcome this problem. The original concept was that all apples arriving on the warehouse loading dock would be sorted and stored by size and grade. Thus, if an order came in for 100 boxes of Red Delicious Extra Fancy, size 113s, only a few bins of that size would need to be removed from storage and run across the packing line. However, the expected growth in consumer packs of fresh apples did not occur. In addition, buyers increasingly ordered mixed loads, that is, carlots made up of a number of different varieties, grades and sizes of apples. Presizing equipment was also relatively expensive because it had to be geared to handling a very large volume of fruit at harvest.

As a result, presizing units have been introduced only in the larger warehouses. Some smaller warehouses can use the facilities of a large neighbor on a fee basis. However, presizing has become a management tool for helping to balance the need for assigning fruit to the appropriate storage and rapidly meeting specific customer orders. Many smaller warehouses have been able to manage their inventory effectively without it.

In a typical presize system, orchard run fruit is dumped and moved through a water flume to the grading tables, graded, mechanically sized, and then returned by water flume to be deposited by size and grade in bins (Schotzko, 1981). The sorted fruit may then be run across an adjacent packing line for packing or be returned to storage until required. The size of the presize unit determines the number of dumping units, the size or number of flumes, the capacity of mechanical sizer and the number and size of sorting tables needed. Product may also be moved through the presizer by mechanical belts rather than water flumes. While water movement tends to cause less bruising, availability of water, sanitation control and disposal of waste water can be a problem in some areas.

STORAGE TECHNOLOGY

Storage of fresh apples requires considerable managerial and technical skill. However, its biggest payoff results from its ability to extend marketing options for growers and packers. A fresh apple, once harvested and left at ambient temperature, will deteriorate at a linear rate. It will become soft, dry and shrivelled. At some time in prehistory, growers discovered the benefits of natural cooling for extending shelf life. Growers used caves or cellars, and used straw or other materials to shield the fruit from air or temperature fluctuations.

The widespread availability of electricity in the 1930s and 1940s, and advances in electrical refrigeration, made it possible for apple warehouses to maintain cold storage facilities at a uniform temperature of 31° F. Smock (1949) at Cornell University, Ithaca, New York, first developed a system which controlled for humidity and carbon dioxide as well as temperature. This system, which has since seen many refinements, became known as Controlled Atmosphere storage (CA for short).

CA has transformed the marketing system for many apple varieties. In the early 1950s, most Red Delicious had to be sold within four months of harvest, usually at a steadily declining price, because of loss of firmness. By the early 1970s, pioneering warehouses were successfully holding Red Delicious up to the following harvest. By the early 1980s, growers were able to market uniformly high quantities of Red Delicious each month between October and June, and very substantial volumes in July and August. This temporal expansion of the market helped Red Delicious to become a major apple variety, especially in North America. It also decimated the market for apples like Winesap, with good natural storage properties but less consumer appeal.

CA technology has been applied with greater difficulty, but with much success, to McIntosh, Golden Delicious, Granny Smith, and Empire, among others. Over time, scientists have learned the advantages of rapid filling of CA rooms, of different combinations of CO_2 and oxygen, of humidity control, and of the use of ethylene for storage of Empire. Systems for ensuring the desired atmosphere have become more efficient, and automated electronic control and monitoring have become commonplace. The technology for fine tuning the quality of each variety stored continues to move forward.

CA storage has brought increased problems of management and quality control. Warehouses have had to become much more cognizant of the timing of harvest and conditions of fruit that is to be stored for different time periods. In general, fully ripe fruit should not be held for late season, but will be of excellent eating quality near to harvest. Apples for late season storage need to be harvested at the preclimacteric stage if they are to be of good eating quality many months later. However, the window for such fruit is quite short. Immature fruit, picked too early, will not ripen into good eating condition.

The longer fruit is stored, the more susceptible it becomes to one or more natural enemies such as scald, rot, or funguses. It will need to be treated with chemicals to reduce the incidence of such problems. However, if treatment is inadequate, or breakdowns occur in refrigeration or atmosphere control systems, entire sealed rooms of fruit can be lost. While storage systems have steadily improved in their ability to deliver firm fruit after prolonged periods, they have not been able to preserve the taste or aroma of fresh-picked fruit. Moisture loss and resultant shrinkage also continues to be a problem.

Storage operations for fresh apples involve a shell building or buildings, equipment to provide the required temperature and atmosphere changes and, for CA, smaller sealed units in which uniform lots of fruit can be stored for a prolonged period. Managers must decide how much total capacity is required for regular storage of loose fruit, for controlled atmosphere storage of fruit for later marketing, and for short-term storage of packed fruit. In general, fruit of good quality can only be marketed out of regular storage for three to four months. The more a warehouse wishes to pack and sell thereafter, the more CA storage capacity it needs. Normally, packed fruit storage will involve less than one week of customer needs.

A second major question for CA storage is the number and size of separate sealed rooms that is desired. The larger each separate room, the longer it may take to fill the room and get it to an optimal atmosphere before sealing. In the event of equipment failure, large financial losses can be incurred. Smaller rooms can be filled more rapidly and the atmosphere tailored more specifically to the needs of particular lots of fruit. However, managing, scheduling, and monitoring demands increase. While the typical CA room currently can hold about 2000 bins (about 800 metric tons) of fresh apples, the size of rooms is becoming much more variable.

PACKING TECHNOLOGY

At some point, either directly after it has arrived from the orchard, after presizing, or after a period in storage, apples must be run across a packing line to make them market ready. The sequence of operations is fairly standard, although the type and configuration of equipment used may vary (Figure 4.1). Generally, the apples are dumped from loose bins into a flume or onto a mechanical belt, through a washer and dryer, over a sorting table where apples are placed on appropriate mechanical belts by grade or placed in a cull shoot. Apples then pass over a sizer from which they are ejected into an appropriate packing bin. From there they are usually packed by hand onto trays or bags and into shipping containers, then placed back on a mechanical belt for sealing, marking and enumerating. Usually the sealed boxes are then stacked into pallet loads for immediate shipment or for storage. Most apples in the Pacific Northwest of the United States are waxed after being washed, but that practice remains controversial in many markets. Cull apples may be additionally sorted for peeling or slicing (the larger apples), for crushing into juice or for disposal as waste.

Packing line systems need to be engineered to fit the total volume of fruit and range in fruit quality that the warehouse expects to meet. For example, a warehouse that wishes to ship packable fruit for a 40-week season and to operate five eight-hour shifts per week can easily estimate how much daily packing capacity it needs to handle all its stored fruit. However, that total capacity can be met by one large packing line or a number of smaller packing lines. A warehouse handling two varieties, such as Red Delicious and Golden Delicious, with widely differing sensitivity to line speed and bruising, would normally handle each on separate lines. However, more machine time will be needed to handle 100 bins of Golden Delicious than 100 bins of Red Delicious. The number and range of grades to be handled will also affect line capacity. For example, a line may be engineered to handle two grades, Extra Fancy and Fancy, where the normal share of packout is 80 percent Extra Fancy, 20 percent Fancy. If the Fancy share of the packout rises to 30 percent, the same volume will take 50 percent longer to run across the packing line. The entire line must adjust to its slowest unit.

Similarly, bottlenecks can occur at any stage of packing. If washing is delayed by dirty fruit or sorting by an unusual level of culls, packers and others may be underutilized. Conversely, if fruit is heavily weighted to certain sizes, some packers may be overworked while others are underutilized. The line manager plays a critical role

FIGURE 4.1. Typical Apple Packing Operations

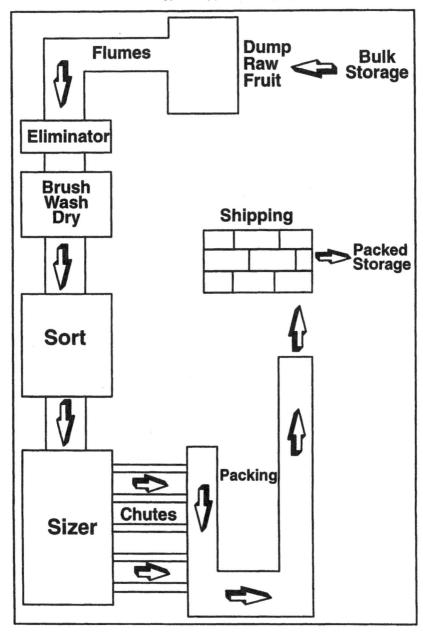

in maintaining a balance in the workload of each piece of equipment and each worker. At the same time, that manager must balance the need for more efficient throughput with the imperative of maintaining fruit quality.

Almost all packing lines require a combination of labor and equipment to get the job done. Labor has tended to be replaced first in those operations which required moving large volumes of fruit or packing materials (such as dumping or moving apples through the system or moving materials to the appropriate packing stations) or involved repetitive, uniform activities such as washing and waxing. Labor has tended to be retained longest in those activities where human judgement is important, such as controlling the rate of dumping, sorting or packing fruit, or checking output. However, many packing line operations have been amenable to automation. Simple replacement of human by machine has been furthered by the development of servo-mechanisms which permitted a machine to adjust or halt its activity until it received certain signals that the next stage in the system was ready to receive product. The development of electronic and computing technology has permitted automatic machines to become more "intelligent" and allowed more complex on-line activities such as sorting and packing to be gradually replaced by machines. The various mechanical operations of a packing line can now be controlled by a single operator at a computer console.

A number of factors influence the mix of labor and equipment in any particular packing line. First is the availability, cost and reliability of labor. In apple producing areas near major commercial centers, labor may be attracted away to other activities. As the cost and frequency of training new employees mounts, automation of operations becomes more attractive. However, there are a number of offsetting constraints to replacing labor with machinery. One is the cost and availability of capital to purchase the new equipment. Another is the teething problems that have arisen in replacing human judgement with machines. Apples, unlike nonliving products, can be easily damaged by the normal bumps, drops and rolls encountered on a fast-moving belt. Possible cost savings from automation have to be weighed against loss in the value of the packed fruit. In addition, electronic devices for sorting by color, firmness or other characteristics have not yet won the trust of many packing line operators.

In theory, if the devices now being proposed come to fruition, fresh apples will be able to be sorted and packed without intervention by any human but the person monitoring the computer controls. In practice, the number of workers required per box sorted or packed has declined little over time. For example, Greig and O'Rourke (1972) estimated that in Washington in 1971, a packing line crew of 56 persons could handle 200 bins (about 3,400 packed boxes of 42 pounds) in one eight-hour shift. Schotzko, (1981) estimated that a packing line crew of 54 3/4 persons could handle 200 bins (about 3540 packed boxes) in the same time. Thus, there appeared to have been little improvement in labor productivity in the decade 1971-81.

The major thrust of automation in the 1970s and 1980s has been in machines which lessen manual labor, such as bag-fillers, tray-fillers, automatic packers, automatic palletizers, etc. However, the greater sophistication of consumers, lengthier storage times and longer distances to market are going to require a greater emphasis on machinery and equipment which enhances the quality character-istics of the packed fruit. For example, using an instrumented sphere, an apple-shaped device loaded with electronic sensors, combined with a movie camera and computer reading device, Hyde, Zhang, and Cavalieri (1990) have been able to identify the number and severity of bruises caused by the structure of the packing line and the flow of apples. Peleg, Klepper, Cavalieri, Pitts and others have been attempting to develop nondestructive tests for firmness, watercore and other internal fruit conditions (Cavalieri and Pitts, 1991). This capability will be incorporated into commercial pack-ing lines in the 1990s. The new technology will be quality augment-ing, but not necessarily labor saving. Rather, it will tend to replace unskilled or semi-skilled labor with labor capable of calibrating and monitoring electronic equipment.

The decision on the appropriate mix of labor and equipment is also affected by the type and quality of pack which the warehouse seeks to produce. At one extreme, a warehouse may simply ship apples of uniform size and grade in field bins for display in bulk at a warehouse retail store or other discounter. At the other extreme, a warehouse may handle a number of different varieties, grades and sizes; packed in consumer packs, bags, jumble packs, tray packs or cell packs; in half-cartons or full cartons; for domestic market or export. Individual apples may be stickered, wrapped or otherwise treated specially. Many warehouses will pack regular fruit, premium fruit and distressed fruit under different labels. They may use different labels for domestic packs and for export packs. Many warehouses use "heavy" packs for export. The extra weight is intended both to compensate for possible shrinkage in transit and as an implicit price discount, since prices are normally quoted per standard 42-pound (19 kilogram) carton. In each case, the number of pack options is strongly influenced by the quality of fruit a particular warehouse assembles and by the expectations of the customers it serves.

EMERGING ISSUES

While many packing operations have, in the past, been set up for convenience on or near to the supplying orchards, many issues other than grower convenience are becoming important to the siting and operation of fresh apple warehouses. First is the availability of an adequate supply of clean water for transporting or washing what is a food product subject to increasingly stricter laws on sanitation and food safety. Next, the sewer systems must be capable of handling the waste water from the plant. Where chemical dips are used prior to storage, there must be adequate facilities for receiving or treating the used dips. Electrical and plumbing facilities must meet local and national codes. There must be adequate ventilation, fire escapes, rest rooms and other amenities for workers. Machines must have adequate worker protection, moving vehicles such as forklift trucks must have adequate warning systems. Walls, floors and working surfaces must be kept free of chemical, bacterial or other contaminants.

While these considerations are imposed on apple warehouses for the broader public good, some warehouses may take on additional commitments to improve their service to growers or customers. For example, many warehouses employ horticultural fieldmen (or women) to work closely with growers year round. Fieldmen ensure that growers produce fruit with desirable characteristics, harvest in synch with the warehouse's needs for meeting different lengths of storage, and remove as much unmarketable fruit as possible before delivery to the warehouse.

A result of the concern for quality in the orchard has been the increasing centralization of quality monitoring in the warehouse under a quality control manager. The more progressive warehouses recognize that while apple warehousing is an industrial process in which cost and efficiency are important, the product is a living organism which must be treated with tender loving care if the final customer is to be satisfied.

A third area in which the more progressive warehouses have taken the initiative is in the use of their accounting staff to analyze the relationship of fruit from specific blocks to quality packouts and to eventual returns. Computer analysis can also be used to continually monitor the performance of the storage and packing operations within plant.

The warehousing and packing industry has experienced almost continual change since the 1960s. It has adapted to the widespread use of wooden field bins for harvest and storage and the replacement of wooden packed boxes by fiberboard cartons. Through increases in size and efficiency, and the addition of sophisticated storage facilities, warehouses have been able to supply packed fruit in acceptable condition twelve months a year with no increase in real cost per packed box. They have incorporated mechanical replacements for labor, automated some operations, introduced electronic devices and experimented with many applications of the computer. The introduction of water flumes and increasing use of forklift trucks has speeded the movement of apples and materials within the warehouse. Palletization has become standard in storage of materials and packed boxes and in loading for shipment. Advances in storage technology have extended the period of storage and marketing for many different varieties.

However, many warehouses continue to have problems in meeting their dual role of servicing grower suppliers and buyer customers. Growers are more concerned about how their products and accounts are handled and about their return per bin of orchard run fruit relative to that received by neighboring growers from competing warehouses. If it sets quality standards too high, a warehouse risks losing growers to competing organizations. On the other hand, the return a warehouse receives for packed fruit is affected both by its cumulative reputation among buyers and by what it puts into each current pack it delivers. Buyers compare warehouses on the value received, that is, on the quality of fruit they receive at a given price. Thus, the short-term goal of keeping grower suppliers happy can sometimes conflict with the long-term goal of keeping customers happy, maintaining a brand reputation or even maintaining the reputation of a producing province, state or country. Often the most market-oriented warehousing and packing operations have the most problems with grower suppliers, unless they make continuing efforts to persuade growers of the long-term payoff from strict quality controls and market discipline.

Finally, as organizations applying industrial processes to living organisms, apple warehouses are going to be intimately affected by the breakthroughs in biological sensors that are being developed for human and animal medicine and for crops. For example, sensors that can detect the current rate of respiration of an apple can be used to trigger an appropriate adjustment in the storage atmosphere. Devices that can peer inside an apple (either literally or metaphorically) can be used to segregate fruit by internal quality characteristics. To meet these changes, new on-line equipment will have to be developed and tested, workers will have to be trained in different skills, growers will have to meet new production standards and new marketing options will be opened up. Thus, warehousing and packing is likely to remain a pivotal part of the apple marketing system.

Chapter 5

Warehousing and Packing: Economic Aspects

Warehousing and packing firms provide a service to apple producers, that of transforming orchard run fruit into a market-ready product. Fresh packers add value just as surely as do processors of applesauce or apple juice, although that added value is often not recognized. Just like the fruit processor, the fresh-pack warehouse aims to cover all costs and leave a margin for return on capital invested. To the grower, the warehousing and packing service is like any other service purchased, such as custom harvesting or aerial spraying. It must be paid for out of the revenues from sales of fruit. The cost of the service must be weighed against the effect on revenues from the manner in which the service is carried out.

Thus, the economics of warehousing and packing are of relevance both to those engaged in the business and to grower suppliers. It can help growers decide whether to set up their own warehousing and packing operation, how to choose between alternative warehouses competing for supplies, and what are reasonable charges for the services rendered. It can help packers decide on the size, location and makeup of an appropriate facility and on the desirability of having one or more facilities.

As an industrial process applied to a living raw material, warehousing and packing seeks to minimize costs subject to two constraints. Minimal quality standards must be met and output must be tailored to the needs of the marketplace. Efficiency alone cannot be used to reduce costs. For example, requiring the same number of sorters to grade more fruit per hour will at some point lead to greater misgrading and reduction in the value of the packed product. Moving fruit over a given machine more rapidly will eventually

lead to an increase in bruise damage. This will be the more perni-
cious because the bruises will not become obvious until some time
later when they are already sealed in a container.

MAJOR OPERATIONS

Since an integrated warehousing and packing facility is an amal-
gam of a number of different operations, it is useful to look at the
economic principles influencing the individual operations before
attempting to look at firm economics as a whole. It is also useful to
look at costs which are fixed in each operation, that is, costs that
will be incurred regardless of the extent of use, and those that are
variable, generally varying in line with the volume of fruit handled.
Finally, it is instructive to separate costs into those that are over-
head-related, and those that are incurred for plant and equipment,
labor, and materials, including fruit.

The assembly operation generally includes loading harvested
fruit in the orchard onto a road vehicle, hauling that fruit to the
warehouse and unloading it at the warehouse. The major elements
of cost in the operation are the equipment and labor used to load and
unload the road vehicle, and the cost of the vehicle, a driver and, if
necessary, a helper on the road vehicle. For example, if a forklift is
used to load or unload bins on a flatbed truck, costs will include the
wages of the forklift driver, depreciation on the forklift, fuel and oil
for the forklift, and other maintenance charges. Overall, loading and
unloading costs should vary only by the efficiency of the forklift
driver.

The on-road costs will vary by the time or distance taken to travel
from the orchard to the warehouse. Labor costs, depreciation charge
(if on an hourly basis) and fuel costs will vary almost directly with
time and/or distance travelled. Use of a straddle carrier can simplify
loading and unloading, but on-road costs will still vary with time
travelled. Also, costs of a straddle carrier will all have to be charged
to the assembly operation, while a multipurpose vehicle's cost can be
shared among other uses. The implication of this for total assembly
costs will be that the average cost per bin assembled will rise as an
individual warehouse increases the distance from which it hauls fruit.

For example, Bressler and King (1970) show that average loading and unloading costs will remain the same but that hauling costs will rise linearly with road distance. Depending on the hauling cost per mile, at some point it will become uneconomic for warehouses to obtain greater supplies of fruit by hauling them from greater distances. In practice, most fruit is grown within 20 miles of its warehouse. Hauling does occur up to 100 miles, often for fruit owned by the warehouse operator. However, beyond that distance, risk of quality deterioration, as well as economics, discourages hauling.

The storage operation at the warehouse is quite different in character from the assembly operation. Normally, the storage operation will require a large storage building containing several smaller, separate storage units and a temperature or atmosphere control system. Annual costs of owning and maintaining these facilities will be incurred even if no fruit is stored. Thus, the nearer to capacity these facilities are used, the lower the average fixed costs per bin or box. Loading and unloading costs will tend to vary directly with the volume stored. The marginal cost of electricity, natural gas, refrigerants or other materials will also be quite small but will vary with the length of time a particular storage area is in use. The net effect will be that, unlike assembly costs, the average total cost of storage per bin will tend to decline as volume increases. However, the addition of a second storage building with accompanying rooms and equipment can cause average costs to ratchet upwards again.

Costs of the presizing operation and the packing operation will share similar elements. Both will involve large buildings and specialized equipment whose annual costs must be met whether or not any fruit is run over the lines. Unit costs will decline as annual volume increases. Both will require labor to tend key stages in the operation, but particularly to carry out the sorting function. Some variation in crew size will be possible in response to rate of flow and quality of the fruit, but unit costs will tend to decline as the volume per shift increases. Both presizing and packing will involve operating costs such as utilities and repairs which will tend to vary with hours of operation. However, the packing line will involve two additional major costs, packing labor and packing materials. Packing labor is usually paid on a piece-rate basis, while packing materials are purchased at a fixed price. Thus, the costs per packed box for labor and materials will tend to be constant. The major source of economies will be in the greater use of the fixed plant and equipment.

In theory, the greater the volume of fruit that can be moved through a plant in a season, the lower the unit cost of packing should be. A plant could be operated for three eight-hour shifts, seven days a week, for 50 weeks. Thus a packing line with a capacity of 5000 packed boxes per eight hour shift could pack 5.25 million boxes in a season. The average fixed costs would be one-third of those incurred by operating the plant for one shift per day. However, few packing operations attempt to run multiple shifts on a regular basis.

There are a number of reasons why multiple shifts have not become common. First is that in developing their packing plant, most planners do not consider multiple shifts as desirable. They build a plant which will be large enough to handle the expected volume of fruit when run during the traditional eight-hour day. Packing operations must adjust to the ebb and flow of market demand. Most operators prefer to pack to orders or anticipated orders and to minimize the volume of packed fruit in storage. If demand is slow, they will close down the packing line for days or weeks. In some areas, it may be difficult to find workers to handle a second shift, and even more difficult for the third or graveyard shift. Supervisory costs will also rise. Managers may be even more reluctant to work evening or night shifts. Labor costs may rise if overtime rates have to be paid. Repairs and maintenance costs will tend to rise. There may be more frequent down time for emergency repairs. Productivity may decrease and the possibility of accidents and pilferage increase. Thus, decreases in fixed costs from running multiple shifts may be offset somewhat by increases in other costs. However, as apple packing lines become more automated and more of the packing line costs become fixed, the economics of multiple shifts may become more pressing.

OVERHEAD AND LABOR COSTS

Overhead costs are costs which cannot be assigned to any of the specific plant operations. They include the costs of general management, secretarial and clerical staff, property taxes, utilities and other costs related to the general management function. As in all economic activities, overhead costs can balloon rapidly if not tightly controlled. On the other hand, certain increases in overhead expenditure can have a major impact on the overall profitability of the operation. For example, the personnel manager can have a major influence on the quality of labor hired, on training provided, on retention rates and on relationships with labor unions. The number and caliber of fieldmen employed to work with growers also can be crucial to the quality of fruit later delivered to the warehouses. Indeed, larger warehouses are increasingly designating one or more officers to be responsible for fruit quality control throughout the operation. Thus, overhead costs can be either a burden or the leaven that makes the whole operation work more effectively.

Labor costs remain a critical part of total warehousing and packing costs. A minimum crew is needed to have the presizing or packing lines in operation. For example, Schotzko (1981) estimated the conventional packing systems required over 30 workers per shift per 100 bins (87,500 pounds) handled. Combined presize and packing systems required about 25 workers per 100 bins. Thus, many large warehouses will employ 150 to 200 workers, most of them involved on the packing line. The wages, fringe benefits, and bonuses received by these workers is the more obvious aspect of labor costs. The less obvious, but equally critical, cost relates to the manner in which the fruit is handled at each stage. For example, rough handling during loading, unloading, or hauling to the plant, in moving product into and out of storage, in dumping loose fruit on the packing line, in palletizing packed boxes, or in loading for shipment to market can have a significant influence on the quality and value of the product delivered.

The skill of the crew on the packing line also has a major influence on the quality pack produced. For example, the line crew can determine the appropriate flow of fruit given the quality being run. They can make adjustments for any unusual situations arising. The sorting crew is particularly critical to the quality of the pack produced. They can determine what shapes are acceptable, what shades of color or striping are appropriate, what size of bruise or defect should be of concern. The physical features of the sorting area can add to the effectiveness of the sorting operation. For example, sorting tables and chairs need to be adjusted to a height where the sorters can stay alert with minimal fatigue. Lighting placement and intensity is also important in accurate sorting. In cases where electronic sorting is also being used, the manual sorters must be alert to the sizes, shapes, or coloring of apples that are most difficult for the automatic devices to detect. Human and machine need to develop a partnership to ensure precise sorting.

The packing crew is also critical in adding value to the final product. Since they are usually paid a piece rate, they have an obvious incentive to increase their take-home pay by working rapidly. However, as the apples are handled and placed in trays, as the trays are stacked, and as the packed boxes are shunted back on the belt, the packing supervisor must be alert that speed does not result in fruit damage. Overfilled cartons, rough sealing, or rough handling in pallets can also reduce apple quality. The warehouse may need to incur overhead costs such as more frequent in-house inspections, or more pervasive quality control programs, to fine-tune its operation. It must weigh the desirability of increasing labor productivity against the customer's demand for a high quality product.

AUTOMATION

Dissatisfaction with the quality of labor and rising wages and advances in technology have encouraged many warehouses to attempt to replace labor with equipment. That equipment ranges from manually-controlled mechanical devices such as forklift trucks or dumping mechanisms, to automated systems such as weight-sizers or tray-stackers, to devices connected to a master control. An electronic reading is flashed to the master control which interprets the reading and sends back a signal for one or more operations to be carried out. Color sorters fall into that category, and many more "intelligent" devices are being developed, either to replace human skills or, in some cases, such as X-ray or ultrasound systems, to go beyond human skills.

The economics of adding a new piece of equipment in the warehouse is rarely simple. For example, a manual-only line of a certain size may require 10 multipurpose human sorters. If an electronic color sorter is introduced, five human sorters may still be required to remove small, misshapen, or watercored fruit. Supervision and training of the human sorters will continue to be required. The cost of the new equipment will include the downpayment, repayment of principal and interest on borrowed funds, installation costs, repairs and maintenance, and the cost of the skilled labor needed to calibrate and monitor the equipment. These costs will have to be charged against the future packs that the equipment will sort during its useful life. The cost per packed box can then be compared with the wage costs saved from reducing the number of sorters. However, the comparison cannot end there unless the electronic equipment leaves the quality of the pack unchanged. If the new system improves quality, the estimated revenue gain can be used to offset the cost of the new equipment. If the value of the product is expected to fall, that must be set off against the expected labor savings.

One other consideration can influence the decision to introduce new equipment. If competing warehouses are using the possession of a particular device as a sales tool either to woo prospective grower suppliers or to earn the goodwill of customers, warehouses may also have to weigh the competitive advantages of possessing a similar piece of equipment. A shiny new piece of equipment with electronic gadgetry can be perceived as a symbol of a progressive operation.

Since the early 1990s, it has been technically possible to automate the entire packing line operation, although not all the available devices can compete in quality control with a skilled human operator. With rapid advances in electronics, computers and biosensors, more and more of the routine or laborious activities will be taken over by machines. In addition, devices will offer packers sophistication in quality control not previously available. The net economic effect will be to replace variable labor costs with fixed capital costs. In this situation, full utilization of the plant's capacity will become increasingly important. Essentially, labor costs are incurred only when the line is operating. If there is a short crop or market demand is slow, so that fewer shifts per season are required, packing line labor costs will also decline. However, equipment costs must be borne, whether or not the plant is in operation. Thus, future apple packing plants are likely to be more highly leveraged (that is, have a greater ratio of debt to equity) and to be more vulnerable to financial loss from declines in capacity utilization. In a short crop year, revenue can decline sharply while annual fixed costs will remain the same. Highly automated, single product, single plant operators will be particularly vulnerable.

MATERIALS

The most overlooked factor in the economics of apple packing is materials other than fruit. Included are items such as waxes, postharvest chemicals, trays, wraps, cartons, pallets, straps, plastic bags, consumer packs, and other miscellaneous items. Materials can often account for one third of total warehousing, storing, and packing costs. For items such as pallets or waxes which are standard throughout the operation, a larger plant can get considerable discounts for volume purchases. However, because of the diversity of the apples delivered to a typical warehouse, even a small plant is forced to differentiate packs by size. The market, in turn, may demand differentiation between bagged and traypack apples, between wrapped and unwrapped, between consumer and wholesale packs. A small plant may have to order as many different kinds of packing materials as a large plant. Even a large plant handling two million packed boxes a year will not be able to get the sort of discounts that can be obtained by a processor of canned soups or other more homogenous food products.

Plants can attempt to gain economies in purchasing by joining a buying cooperative or by pooling their orders for materials. This is likely to be most feasible and most sustainable for items such as pallets or waxes that cannot be used as a merchandising vehicle. However, paper wraps, plastic bags, apple stickers, and outer cartons are all frequently used to carry the logo and the advertising message of the individual warehouse. The warehouse must weigh the benefits of maintaining brand recognition in all of its materials against the savings from buying materials without brand identification.

A variant of the same problem is whether to identify all materials with the firm's own brand or with a logo supplied by an umbrella organization such as the Washington Apple Commission or the Cape (South African) apples. While these umbrella organizations can offer various economies in purchasing, individual warehouses are often reluctant to display anything but their own brand on materials that are on display in wholesale or retail markets.

Materials use is likely to change in response to consumer and business concerns about waste disposal. Paper products utilize scarce forest resources and involve controversial treatment processes but are biodegradable under appropriate conditions. In contrast, petroleum-based plastic products come from a nonrenewable source and are not easily biodegradable. New materials that are biodegradable and new methods of biodegrading existing materials will change the choices available to warehouses, but will also change the cost implications.

FRUIT QUALITY

Perhaps the factor that has the biggest effect on day-to-day fluctuations in apple warehousing and packing costs is the quality of the fruit passing over the packing line. Greig and O'Rourke (1972) found, in a study of U.S. and Canadian apple packing plants, that the percentage of top grade (Extra Fancy) apples varied from 43.2 to 85.4 percent for Red Delicious and 38.9 to 67.5 percent for Golden Delicious. The average size of apple ranged from 90.7 to 143.1 per box for Red Delicious and 99.9 to 128.9 for Golden Delicious. The percentage of dumped fruit culled ranged from 9.2 to 21.1 for Red Delicious and 17.4 to 58.7 for Golden Delicious. The less homogenous the fruit dumped on the packing line and the higher the number of defects, the more difficult the sorting process becomes. The smaller the apples, the more must be handled to fill a standard carton manually. Thus, lower quality or smaller size fruit both cost more to handle directly. In addition, slowing of throughput increases the average fixed costs per box. This presupposes that the market rewards packers who segregate distinctly by grade and size. However, combination packs of mixed grades or sizes are still available from various sources.

Controlling the costs associated with low quality fruit can be achieved in a number of ways. Careful selection of growers from whom a warehouse draws is an important first step. Many warehouses also employ fieldmen to work with growers throughout the year to ensure delivery of appropriate quality at harvest. However, despite the best efforts of all concerned, mother nature occasionally provides a crop with exceptional internal or external quality problems. The earlier in the warehousing system that fruit not suitable for fresh packing can be eliminated from the fresh system, the better. Selling orchard-run fruit directly to the processor, sampling at the unloading dock, elimination on the presizing line at harvest, and separate storage of marginal product, can all contribute to lower warehousing and packing costs and to better quality packs. Depending on the firm's normal operations, these steps may not involve any additional expenditures. However, they will almost always lower packing costs and reduce storage losses.

LONG-TERM AND SHORT-TERM CONSIDERATIONS

It is clear from the foregoing that no one financial analysis can capture the nuances of costs in apple warehousing and packing. Depending on the length of planning period involved, the proportion of factors that are considered fixed can change. In addition, the business decisions which will be posed will vary with the time period covered. For example, in projecting finances for a ten-year planning period, a warehouse can alter the amount of storage space, the number and size of presize and packing lines, and the mix of machinery and labor utilized. In this situation, management is primarily interested in the return on investment from the facilities and equipment selected. That will depend heavily on the volume of apples handled in each year. Any given acreage will generate a wide range of production due solely to weather conditions in any season. If the warehouse expands its facilities to handle the maximum possible crop, not only will its capital needs be at a maximum, but it will also know that in many seasons the plant will be operating below full capacity. The warehouse will have to adjust its charges to cover those years of reduced crops when average fixed costs will be above the minimum.

At the other extreme, if the warehouse only has sufficient facilities to handle the minimum crop, in many years it will have to turn away some of its regular growers' fruit. In some cases, it may be possible to contract with a neighboring warehouse to handle the excess. However, in many cases, weather factors will cause all warehouses in a region to have large or small crops simultaneously. The cumulative effect is that all warehouses tend to plan for their maximum crop and that the industry, in general, tends to have excess capacity in most years. That excess capacity has to be paid for by charges on the fruit actually handled.

A major issue in long-term financing will be the share of capital provided from debt and from equity. Debt, whether received from banks or other financial institutions, will require annual repayments including both principal and interest. The higher the proportion of debt, the higher the interest rate, and the shorter the loan terms, the greater will be the burden of debt to be charged against the apples handled. All these factors will have to be weighed in any firm decision on plant or line expansion. Equity capital tends to be rewarded from the profits remaining after debt is serviced. Thus, equity holders may be willing to accept minimal rewards for one or more years. However, in the long run, equity will not be attracted to or retained in the apple industry if it fails to generate dividends comparable to those available in alternative investment opportunities.

Investors in warehousing facilities tend to face less fluctuations in revenue and profitability than do growers of a comparable volume of apples. An example of a group of growers who supply a single warehouse will help to illustrate. Suppose that, on average, these growers deliver one million boxes of packable fruit to the warehouse. Their average cost of production is $6 U.S. per packed box, including a return to family labor, management and capital. The average warehousing, storing and packing cost (including a normal return on family labor, management and all capital) is $6 per packed box. A shipping point price of $12 per packed box will yield the growers combined $6 million and the packinghouse $6 million. In a region-wide short crop year, if growers deliver only 750,000 packed boxes, the likelihood is that shipping point price will rise by 25 percent to $15 per packed box. If, as is customary, the warehouse maintains its normal $6 per packed box charge, the warehouse will get revenues of $4.5 million and the growers $6.75 million. In a region-wide large crop year, if growers deliver 1,250,00 packed boxes, the likelihood is that shipping point price will fall by 25 percent to $9 per packed box. If the warehouse maintains its normal $6 per packed box charge, the warehouse will get revenues of $7.5 million and the growers $3.75 million. Thus, while warehouse revenue will rise 25 percent with the large crop and fall 25 percent with the short crop, grower revenue will move in the opposite direction. It will rise with the short crop (but only by 12.5 percent) and will fall with the large crop (but, more dramatically) by 37.5 percent. For the three years combined, warehouse revenue will average $6 million (that is, provide a normal profit) while grower revenue will average $5.5 million (below full cost).

The above example illustrates well the inherent conflict between growers and packinghouses with respect to volume control. The nearer to maximum capacity that a packinghouse can operate, the greater its total revenue, the lower its average cost per packed box, and the greater its profits. In contrast, growers generally find that above a certain level of supplies, their price per packed box drops more rapidly than do their costs due to improved economies of scale. In the example above, they might be able to lower average production costs in a large crop year to $5 per box. That would still mean that the crop cost $6.25 million to grow but yielded only $3.75 million in revenue. Packinghouses face the reverse situation in a short crop year with higher than normal average costs and lower total revenue. However, packinghouses have the advantage of being able to set their packing charges at a level which takes account of expected fluctuations in volume between seasons.

The second planning period for which financial analysis is critical is the single crop year or packing season. In the Northern Hemisphere by August, and in the Southern Hemisphere by February of each year, warehouses have reliable estimates of the volume and quality of apples of each variety they can expect to handle in the coming season. They also have in place the complement of buildings, storage rooms, field bins, presize lines, packing lines and vehicles needed to handle the crop. Their first goal is to get the crop as expeditiously as possible off the trees and into cold storage where further quality deterioration can be retarded.

If, because of the pace of harvest, incoming fruit exceeds storage capacity, the warehouse may rent short-term storage to handle the overload. If the overall volume is beyond its capacity, the warehouse may try to rent storage space for the season. Conversely, if its capacity is likely to be underutilized by its own growers, the warehouse may try to lease some of its space to other warehouses. Because of the tendency for crops to be above or below average region-wide in each climatic zone, finding a storage partner is often difficult.

A number of decisions also need to be made with respect to presizing. Ideally, all fruit would be presized at harvest time so it could be stored with other fruit that is similar in size, color, storage life, etc. Annual fixed costs could be shared over all fruit in the warehouse. However, the cost of presizing equipment means that most warehouses choose a presizer which would take 100 or more eight-hour shifts to handle their entire manifest. In theory, with multiple shifts, it might be possible to fit 100 shifts in a prolonged harvest. In practice, few firms are eager to utilize multiple shifts. Accordingly, warehouses must decide which lots of fruit are to be presized at harvest, which lots are to be stored first and later presized, and which lots need not be presized at all. Clearly, the earlier and more precisely fruit can be segregated by its market characteristics (e.g., processed use versus fresh pack, different size counts, solid color versus striped, early storage versus late storage), the more easily the warehouse can pack on demand. On the other hand, any pass of fruit over the presizer will involve some variable costs which must be subsequently recovered from growers. In addition, the more categories in which fruit is segregated, the greater the monitoring and record-keeping costs involved in managing inventory. The additional costs of presizing must be weighed against the improvement in product quality and in service responsiveness required by customers.

STORAGE COSTS

Storage presents different economic issues. The warehouse begins the season with a fixed complement of storage space including refrigerated, controlled atmosphere and packed fruit storage. In general, only a minimal amount of fruit will be held in packed storage, so it will be ignored in subsequent discussions. A major difference between controlled atmosphere and refrigerated storage in the United States is that apples must normally have been sealed in controlled atmosphere storage for 90 days in order to qualify for the CA designation. Thus, CA storage space is normally subdivided into rooms which can be filled rapidly and sealed until needed. Smaller rooms, when opened, can be more easily marketed in a reasonable time so that the quality derived from CA storage is retained.

Warehouses have become increasingly concerned about the speed with which CA rooms are filled, the rapidity with which the desired atmosphere is achieved once the room is sealed, and the precise levels of oxygen, carbon dioxide and relative humidity to be maintained. This means greater management attention to the logistics of having the needed fruit ready to store, and improved technology and greater supervisory effort while the fruit is in storage. Inevitably, this greater precision involves greater costs. Apples can be accumulated in refrigerated storage after harvest so CA rooms can be filled rapidly at the appropriate time. Much depends on what mix of qualities are to be sealed in a single room. The more discrete quality categories that are identified and the more separate rooms that are maintained, the greater the monitoring and management costs.

However, having fewer, larger CA rooms must also be weighed against the financial consequences of fruit loss in larger rooms. Even twenty-four hour automatic monitoring systems can give false readings or fail to warn of a harmful change in temperature, gases or storage diseases. It is not unknown for all the apples in a late CA room to be found to be unmarketable. In a typical warehousing operation, the loss of one 2000-bin CA room can eliminate all profits for the season.

In making an optimal decision on storage rooms for its facility, there is no substitute for careful record keeping and analysis of the history of each room. The output of each room, that is, the quality of marketable fruit, must be compared relative to the quality of the fruit put in, the storage regime imposed and the length of time the fruit was stored. In this way, over a period of years, warehouse management can relate the payoff from different storage facilities and treatments to specific varieties, growers, orchard districts or other factors affecting payoff. Both the average payoff and the variance of the payoff (a measure of risk) can be established.

PACKING COSTS

In the case where a warehouse has multiple packing lines, the same sort of comparative analysis is needed. For example, a packing plant may have a separate line designed to handle Red Delicious apples and another line to handle the more sensitive Golden Delicious variety. Annual supplies or marketing needs for the two varieties will rarely be equal. The warehouse may choose to build two lines with identical physical dimensions, or identical handling capacity per shift. It may build lines of different capacity so as to run the same number of shifts per year, or to meet different expected daily market requirements. Each choice will commit the packinghouse to a given cost structure many years into the future. For example, if the lines have identical physical dimensions, fewer Golden Delicious than Red Delicious can be packed per shift. The Golden Delicious line will have to be used for more shifts per season to generate the same annual output of packed boxes and the same average packing-line fixed cost per packed box. However, running more shifts will increase the annual cost of utilities, repairs, maintenance and shift-related expenses. If the lines have identical packed box output per shift, the Golden Delicious line will have to be larger and thus involve larger fixed costs per box.

The analysis becomes more complicated if a single line is to be used for more than one variety of apple, or, as in some cases, for apples and pears. Since the dimensions of cull chutes or grade lines, the mechanical layout, or the types of surfaces or padding that are optimal for one variety may not be optimal for another, managers must trade off higher costs than necessary for the less sensitive variety against better quality output for the more sensitive variety.

At the other extreme, a warehouse may choose to have a number of specialized packing lines, and/or a number of multipurpose packing lines. Multiple lines provide management with considerable flexibility in allocating lots of fruit to the most suitable line for maximizing quality. They also provide insurance against the breakdown of a single major line. However, in general, the more separate lines, the greater the total annual cost which must be charged against the fruit to be packed. Flexibility comes at a price. Again, only analysis of detailed records over a number of seasons can help a warehouse make the choices most suitable for its circumstances.

IMPORTANCE OF BALANCE

A major overriding factor in warehouse operations will be proper balance between the volume and quality of fruit received, the capacity of refrigerated and controlled atmosphere storage, the presizing line, the packing line or lines and the market demand. Warehouses go to great pains to predict what their likely fruit receipts will be in future years and attempt to build storage facilities and packing lines to meet those needs. The volume of controlled atmosphere storage needed will also depend on the expected distribution of sales throughout the year. A warehouse can err in having too little storage, especially in a large crop year, when it will have to scramble to rent alternative space that may be either less suitable or more costly. On the other hand, it can err in having too much storage, which is only full in a peak year, so that average fixed costs per packed box are driven up in off-peak years. Similarly, its complement of packing lines may be underutilized or may be used so heavily that repair costs rise and breakdowns are common.

Of course, achieving an overall balance in the warehouse operation is easier said than done. It is not unusual in an individual warehouse for fruit supplies in any season to vary 50 percent above or below the five-year average. Crop prediction is particularly difficult for young orchards. Quality and storage life of fruit can vary markedly from year to year. Adding to either storage or packing facilities may require the warehouse to borrow. Its ability to borrow will be affected by its current financial strength, the general availability of credit and the cost of credit (interest rates). However, the need for the warehouse to achieve balance in its long-term operations will remain.

The same need for balance will be found within each shift on the packing line. Just as a chain is as strong as its weakest link, so a sequential operation such as a packing-line is as slow as its slowest activity. For example, the dumper needs to dump fruit at a rate which utilizes the capacity of the line. However, if fruit is moving through too fast it may not be getting properly washed, waxed, or dried. If fruit is of uniformly high quality, the normal complement of sorters may be too many. If fruit has a high proportion of culls and different grades, more than the normal complement of sorters may be needed to avoid slowdown. If fruit is heavily biased to certain sizes, packers may need to be reassigned to those tables. Only by regular analysis of how different lots of fruit are most efficiently handled, can a warehouse fine-tune its operation to handle the maximum packout per shift. Any time fewer apples are packed per shift, or more shifts are required to handle a given volume of fruit, the average cost per packed box will rise.

The costs per packed box will also be affected by the number of packs a warehouse prepares. The more different packs prepared, the more diverse the inventory of cartons, labels, trays, wraps and bags that must be carried. For small runs of different materials, the warehouse is less likely to get purchasing discounts. More materials will make packing-line supply more complicated and costly and increase the storage space and inventory management problems. However, choice of packs must also be influenced by what customers want. The incremental costs of a more diverse pack must be clearly offset by the incremental returns from customers.

INCORPORATING NEW TECHNOLOGY

The last major cost issue that warehouses must face is the decision whether or not to incorporate new technologies into their operations. An important check is to constantly reexamine the goals that management is trying to meet. These goals may include controlling quality, saving energy or other resources, reducing costs, replacing manual labor, enhancing existing operations (e.g., color sorting), adding new operations (e.g., tests of internal quality), improving customer perceptions of one's operations or one's product, or meeting stiffer environmental, safety or health regulations. Next, management needs to ascertain (1) if the technology actually delivers what it promises (e.g., Does it effectively clean waste water?), (2) what the fixed and variable costs of the new technology may be, (3) what the side effects of introducing the new technology may be (e.g., requiring retraining of existing staff or acquisition of new expertise), and (4) what the market effect may be. Hopefully, the new technology will improve market value. However, some new technologies have brought greater efficiency at the cost of greater damage and lower fruit value. In some cases, the firm may have no choice but to introduce a costly new technology in order to meet tighter government regulations.

Almost certainly, the technologies of storage, warehousing and packing will continue to change in the future. The apple industry will continue to select from the best technologies developed for the citrus, tomato, potato and other produce industries. It will adapt processes developed by the computing, electronic, medical and other related industries. Innovators within the apple industry will develop new ways to meet their cost, efficiency and quality goals. And suppliers of plant and equipment to the apple industry will compete in creating improvements (sometimes small, but occasionally dramatic) in their products and services. The warehouses that survive and flourish into the twenty-first century will be those that constantly scan the new technologies for innovations that are most appropriate and beneficial in their circumstances.

Chapter 6

Transportation and Wholesaling

While there are many ways in which fresh apples move from the producing area to consumers, a dominant force in the commercial apple marketing system is wholesalers. In contrast to packing-houses in producing areas which take heterogeneous lots of grow-ers' fruit and combine them into bulk standardized loads, wholesal-ers at the consuming end buy bulk loads and break these down into smaller lots tailored to the specific needs of retail outlets. Packing-houses and their shippers in producing areas are tied to wholesalers in consuming areas by the umbilical cord of transportation. Until recently this distribution system was little known and of little con-cern to the general public. However, growing interest in health and nutrition is bringing increased scrutiny to that system.

In most systems around the world, price of fresh apples is deter-mined in the triangle between the producing area shippers, the consuming area wholesalers, and the transportation system. In North America, domestic apple prices tend to be set by negotiation FOB shipping point. In northern Europe and Asia, price continues to be set at wholesale receiving points.

The term "wholesaler" is used here rather loosely to describe entities which buy in bulk and break down loads into smaller lots for resale. These functions are often performed wholly or partly by firms which call themselves brokers, agents, distributors, exporters, importers, or wholesalers, or some combination of these names. They are also often performed by the buying department, distribu-tion center or wholly-owned wholesaling subsidiary of a retail chain. However, to simplify the discussion, we will refer generi-cally to the wholesaling activities as if they were all conducted within a single independent firm.

The term "shipper" is used to describe the agent in the producing area who trades with the wholesaler in the consuming area. The shipper may draw fruit from a number of smaller packinghouses or may be affiliated solely with one large packinghouse. In either case, the shipper will be the key link between the transportation system and the buying wholesaler.

Shippers, wholesalers, and transportation companies will differ in their goals in handling fresh apples. A mutual understanding of these different goals can go a long way toward reducing conflict in the distribution system.

Wholesalers normally deal in a wide mix of produce items so that their facilities can be used year round. Many also carry a wide array of other food and of nonfood items. In apples, as in other products, they prefer to have a number of alternative sources of supply and healthy competition in price and service between producing regions and firms. They want a quality product and reliable transportation at the lowest possible delivered cost. They are sensitive to final demand in two ways. If product movement at retail is slowed by snow, heat, or other inclement weather, or boosted by holidays or special events, they must be prepared to meet the changing needs by adjusting their inventory. Second, if product consumed is unsatisfactory, that message will be relayed to them very rapidly.

Shippers, in contrast, are usually specialized in one or two fruits such as apples and pears. If possible, they want to become the consistent, preferred supplier to major wholesalers. They, too, want reliable transportation at a competitive cost and delivery of a quality product so the buyer will make repeat purchases. They want the minimum of buyer complaints and of postshipment price adjustments. They are usually concerned about maintaining product movement at some norm so that their inventory is steadily depleted as the season progresses. Above normal inventories make shippers nervous and encourage wholesale buyers to seek price breaks.

The transportation company, like the wholesaler, usually handles many compatible produce items. It seeks to keep its trucks or railcars moving with full payloads whenever possible. To that end, it tries to schedule its vehicles within major produce traffic flows, avoid dead heads, and avoid running empty. Both shippers and wholesalers will be dependent on competition between transportation firms to keep freight rates low and improve service. In contrast, because of their large investment in fixed facilities, transportation companies frequently seek government blessing for managed rates and standardized services.

Competition between shippers and between transportation companies is strongly influenced both by the needs of wholesalers and by the efforts of individual firms to gain a market advantage over their rivals. Wholesaler buyers decide when to place an order, when they need that order delivered, where they need the order delivered and the makeup of the load. Buyers can specify the quality of palletization, placement in the truck or railcar, temperature monitoring required and other features. Both shippers and transportation companies can respond by finding new technologies that can either enhance quality or reduce cost of the same service. A number of large suppliers of transportation equipment have been in the forefront of advancing the technology for transporting perishables.

TRANSPORTATION OF PRODUCE

The dominant carriers of fresh apples over land have been rail and truck. Rail has been able to hold its own in countries where producing areas are far from main consuming areas and highways are unreliable. However, in most developed countries, road conditions have tended to improve as incomes increased. Long-distance haulage by truck is made easier by limited access freeways or motorways on which fully loaded modern trucks can move at over 60 miles (100 kilometers) per hour. Trucks have the added advantage of providing point-to-point service without intermediate loading or unloading as for a conventional railcar. A load of apples can move across the continental United States from Wenatchee, Washington to Miami, Florida, a distance of 3,000 miles, in about three days. As a result of this service advantage, trucks now handle over 90 percent of all fresh apples shipped long distance.

In theory, rail should have a cost advantage over trucks for distances greater than 600 miles. Beyond that point, the higher loading and unloading costs for a train are offset by economies of scale in hauling. A crew of three can control a train hauling one hundred cars, each with 80,000 pounds of fruit. In contrast, each 80,000 pounds of fruit on a truck requires at least one active driver. For a long haul, two drivers may be required. However, the train system requires railroad personnel at stations, crossings, refueling stops and other points which must be charged to individual loads. These services are provided to the trucker at no cost on public highways and freeways. In contrast, truckers pay sizable annual road taxes, which must be allocated to individual loads, and pay tolls on specific routes. In many countries, too, railroads receive both direct and indirect subsidies which conceal the true costs of operation. Thus, it is very difficult to compare actual costs of shipping by rail versus truck.

Despite the foregoing, actual truck charges for hauling apples in the United States tend to rise rather gradually with distance hauled. For example, the average cost per 42-pound packed box in the year 1990 from central Washington State to Los Angeles (a distance of 1050 miles) was $1.60, while the Yakima-New York City cost was $3.37 for a 2770 mile haul (USDA, March 1991). These one way charges do not give a clear picture of the economics of trucking because of the trucker's ability to accept back-hauls, both direct and indirect. For example, after unloading Yakima apples in Maine, the trucker may haul a load of Maine potatoes to Arkansas and then haul a load of Arkansas chickens to Yakima before starting the process all over again.

On average, in 1990, the fixed cost per 42-pound packed box was about 40 cents and the additional cost per mile about one-tenth of a cent (Figure 6.1). Thus, the charge for a haul of 2,000 miles would be expected to be about $2.40 per packed box. Since U.S. fuel costs were low by international standards, truck sizes larger, and traffic congestion less than in many more densely populated nations, transportation charges would be likely to be much higher in many other countries in Europe and Asia. However, the same principles would apply. Transportation costs would consist of a relatively fixed charge for loading and unloading and of a segment that varied with distance travelled and time taken.

In theory, costs for hauling by ship should be similarly dominated by costs incurred while the ship is in port (which are relatively fixed) and costs incurred on the high seas (which are mainly a function of time and distance). However, the quality of service

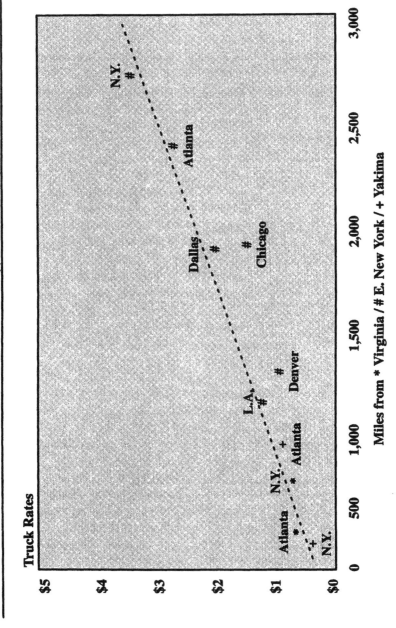

FIGURE 6.1. Truck Rates for Apples, 1990 Average

varies much more widely between ships than between individual trucks or railcars. Shippers may use liners which follow a regular schedule between ports, tramps which move where cargo loads are available, or chartered vessels which can be secured for a specific delivery. Only a large shipping organization can get the economies of scale from filling a chartered vessel. For smaller loads, the shipper usually has to share space with other products. The price charged will be influenced by the value of these other products and what the traffic will bear.

However, perhaps of greater consequence has been the ability of ocean-going carriers to set prices among competitors through a series of rate agreements. Even in the United States, where antitrust laws are enforced relatively vigorously, oceangoing carriers have exemptions from antitrust laws. These rate agreements (or conferences) tend to be negotiated for specific trade zones such as Trans-Pacific Westbound, Asia-North America Eastbound or United States-North European. Many carriers are members of a number of conferences. In a situation such as prevailed through the mid-1980s when imports to North America from Asia greatly exceeded trade in the other direction, ships frequently operated below capacity on the return trip to Asia. This tended to push up the rates for cargo moving to North America and lowered the rates for cargo moving to Asia. No single agricultural commodity was immune from these external influences. Conferences have also been able to negotiate agreements on reducing their capacity in order to firm up rates. The net effect is to limit the options available to agricultural shippers in lowering ocean transportation costs.

The only feasible alternative transportation mode available to shippers across the oceans is air transportation. However, given current technology, airlines can generate many times the revenue from hauling passengers that they can from hauling bulk freight such as fresh produce. Air freight is viable only for items of very high intrinsic value (a category in which apples rarely fall) or in situations where the return to the shipper is sensitive to a one-day delay in arrivals. For example, in early markets for highly seasonal items like fresh sweet cherries or fresh asparagus, prices can sometimes drop 20 to 30 percent in one day. Since fresh apples are now available in large quantities twelve months a year, such price swings are most unlikely. Until the technology of transportation permits shipment of bulk commodities at rates closer to those of cargo ships, fresh apples are not likely to find a viable alternative to cargo ships.

WHOLESALERS OF PRODUCE

The general purpose of a wholesaler of produce is to receive large loads and break them down into smaller volumes to meet the needs of produce retailers. In the mid-nineteenth century, when long distance hauling of produce first became common, wholesalers tended to cluster around major rail terminals such as New York and Chicago, or major port cities such as London or Hamburg. The prevailing clients for wholesalers were small, independent retailers. Indeed, in many cases, such retailers were so small and numerous that terminal wholesalers served many secondary wholesalers who in turn serviced a local group of small retailers. Frequently, price was determined at auction markets in the rail terminal or port district.

While remnants of that system still remain, wholesalers have undergone many transformations to meet changing circumstances. For example, the emergence of retail chains in the late nineteenth century, the development of the supermarket pre-World War II, and the eventual dominance of supermarket chains in the retail food business, forced wholesalers to adapt or die. Many became sponsors of voluntary groups or cooperative groups of small stores in order to offset the financial and marketing muscle of the chains.

In the 1950s and 1960s, A&W, McDonalds, Arby's, Burger King, and Kentucky Fried Chicken brought the chain concept to eating fast food away from home. And in the 1970s and 1980s, they and their many clones catered to an increasingly health-conscious population by adding more fruits and vegetables, including the ubiquitous salad bar. At the same time, the dominance of supermarket chains was being nibbled away by convenience stores and by warehouse food operations. Thus, throughout the food business, both for at home and for away from home eating, the typical retail outlet was more likely to be represented by a chain buyer than to be a small independent operator.

Many of these large retail and restaurant chains have inhouse integrated wholesale operations which are capable of dealing directly with shipping point suppliers for all their produce needs. Many have their own receiving and storage facilities and their own distribution fleets to serve member retailers. They have become the major direct competitors of specialized produce wholesalers. In turn, the specialized produce wholesalers must be extremely flexible both in servicing their regular customers and in serving as supplementary suppliers to chain buyers. Chain buyers can use specialized wholesalers to make up temporary shortfalls in inventory or during periods of unstable prices in order to reduce speculative risk.

Wholesalers have also had to adapt to the tremendous expansion of away from home eating in other outlets such as schools and colleges, airlines, sports events, cruise lines, hospitals and nursing homes, in-plant cafeterias and upscale restaurants. In all of these, the trend towards more healthful eating has brought an increase in the volume and variety of all kinds of produce served to their patrons.

Apples have had relatively mixed fortunes in penetrating these many different outlets. Partly, this reflects a lack of understanding by many in the apple industry of why the wholesale distribution system has changed as it has.

THE CHANGING PRODUCE SYSTEM

McLaughlin and Pierson (1983) carried out a comprehensive analysis of the U.S. produce system in the early 1980s. In a series of in-depth interviews they compared the objectives, strategies and standard operating procedures of participants in the produce distribution system McLaughlin (1983). They concluded that failure of the various participants to understand each other's objectives and operations was leading to increased costs and decreased efficiency in the system. Changes that were occurring were poorly understood. For example, because of higher labor costs at the retail level than at shipping point, there had been a tendency for activities such as consumer packing to be passed back to shipping point. Again, because of the higher cost of real estate at urban receiving points, than at rural shipping points, there was a tendency for the system to prefer to carry inventory at shipping point than at urban wholesale or retail warehouses.

However, such shifts also could have unforeseen effects on the profitability of various parts of the system and on the burden of risk. For example, in the case of fresh apples, preparing consumer packs offered packinghouses an additional opportunity to add value to their products. It also increased the risk that product deterioration in transit would either increase the need for repackaging or decrease the quality and value of the pack. Similarly, the increased demand for storage at shipping point gave the shipper an additional means of adding value and capability of developing alternative marketing strategies. However, it also gave the buyer the opportunity to reduce inventory costs and to pass back risk in times of uncertain demand.

The driving force behind many of these changes identified by McLaughlin and Pierson in the 1980s has become even stronger in the 1990s. Changing consumer preferences, demographics and lifestyles led to further increases in per capita consumption of fresh fruits and vegetables, and an increase in the consumer's demand for greater variety and novelty in fresh produce. Items almost unknown in the 1980s, such as kiwi fruit, star fruit or Granny Smith apples, were, by the 1990s, known to most consumers. Retailers responded by devoting an increasing share of ever larger stores to fresh produce. Keeping their shelves stocked on a 12-month basis required them to draw supplies in season from different parts of the world. Thus, wholesalers in Hamburg or Philadelphia were as likely to be negotiating for fresh apples with suppliers in Chile or New Zealand as in nearby producing regions.

This international extension of the produce system increased the problems identified by McLaughlin and Pierson. For example, Chilean shippers struggled to understand the influence of U.S. marketing orders on the competitive environment for their products, while U.S. suppliers struggled to cope with the targeted programs of the New Zealand Apple and Pear Board. Shippers in general continued to have difficulty coping with the multiproduct pricing systems of wholesalers or retailers, or with the differential marketing strategies of integrated wholesalers (i.e., owned by retail chains) and independent wholesalers which dealt with each retail buyer in arms length transactions.

McLaughlin and Pierson characterized the produce marketing system as fragmented and dispersed. While price was the principal coordinating device, price alone was often unable to carry all the information necessary to coordinate supply and demand. For example, as Al Saffy and O'Rourke (1984) have shown, prices at shipping point may be declining in response to increased supplies, at the same time that retail prices are flat or rising because retail pricing strategy is tied to the selection of produce available not the supply of a particular product. One part of the system, by lowering price, may be signaling buyers to increase purchases while the ultimate buyers (consumers) are receiving a price signal to hold or reduce their level of purchases.

On the other hand, when retailers wish to increase the volume of an item sold, they may lower price dramatically below the level justified by changes at shipping point, expand the display space available, and increase the in-store and media advertising for the item. If consumers stock up during the advertised special, demand may fall off temporarily thereafter as consumer inventories are run down. However, shippers may interpret such temporary lulls in purchasing as more fundamental declines in demand and wrongly reduce price.

This situation results partly from imbalance in information. A supermarket chain has access to detailed daily information on consumer purchases of fresh apples and many competing products. It can also receive publicly provided information on market conditions at many shipping points and wholesale markets. In contrast, the typical shipper cannot find any publicly available information on current market conditions at retail. In addition, shippers themselves frequently oppose the provision of such retail information because of a belief that their private information sources give them a competitive edge over less well-informed competitors. This is the principle that in the Kingdom of the Blind, the one-eyed man is king. A more sober assessment might lead them to conclude that an entire class of market participants–shippers–is disadvantaged because of the greater information available to their trading partners, the large retailers. Wholesalers are less vulnerable to absolute changes in prices than are shippers, because they can pass on such changes to their small retail customers. These, in turn, are able to monitor the vagaries of pricing of their large retail competitors and to tailor their pricing and ordering accordingly. However, it is uncommon to find small retailer price behavior as out of step with shipping point prices as that of large retail chains.

McLaughlin and Pierson also noted problems with the incentive structure facing system participants. Incentives are often conflicting between buyers and merchandisers in the same wholesaler or chain grocery operation, or between retail produce managers in independent stores and affiliated wholesalers. Incentives are often short-run. For example, buyers are rewarded for beating the competition in procuring new season fruit and sellers for being first to market. However, this frequently leads to the shipment of unripe fruit which can potentially alienate customers for the long-run. Few incentives exist for experimentation or innovation. In particular, there is little incentive for firms at one level in the marketing chain to work with firms at other levels to improve the overall efficiency or effectiveness of the system. The more distant the levels, for example, of growers from retailers, the less the understanding and cooperation. In turn, firms in between, such as shippers, brokers or wholesalers, feel little responsibility for strengthening grower-retailer links.

As consumer interest in produce has grown, retailers have become more attentive to consumer needs and less tolerant to shipper practices such as off-quality or out-of-condition products, which failed to meet those needs. Shippers increasingly face a buyer who perceives himself or herself as the consumer's buying agent and not as the shipper's marketing agent. Consumers are dictating what, when, and how they shall receive their produce, not the producer or packer of the produce. This has radical implications for the traditional system of choosing packing methods, box sizes, grades and quality standards. For a long time, these were set by growers and processors or by the public agencies who represented them, such as state or national departments of agriculture.

Technology has been, and continues to be, a major driving force for change. Universal Product Codes and sophisticated computer systems have given large retailers the ability to analyze the sales and profit performance of individual produce items and to discontinue those that fall below established criteria. Many large multiproduct wholesalers now have the same capability. Computers can also be used to tailor inventory to the precise needs of customers. As discussed previously, changes in technology at shipping point are giving suppliers the capability of serving the needs of their customers in many different ways. Opportunity exists for extension of information management methods to every aspect of the produce marketing system.

Together, these forces are leading to major structural shifts in the produce marketing system. Mergers have become commonplace among firms at each level of the system. What at one time is considered to be an economic size, rapidly is deemed too small, leading to further mergers. Multicommodity firms which made their reputations on branded processed products, continue to move into branded produce items. Some have integrated horizontally across international boundaries and vertically from growing and packing to wholesaling and transportation (owning their own fleets of trucks, railcars or ships). As McLaughlin and Pierson pointed out, these multicommodity, multilevel operations are much more apt to think like buyers, to have market power which can to some extent offset the growing power of retailers, and to be able to afford access to good consumer and market information. There is the danger that the produce system is becoming splintered between these giants and the single-region, limited-line shippers whose fate can be decided by a localized setback in a single season.

McLaughlin and Pierson concluded that the produce system existing in the United States in the early 1980s was cracking under the strain of adapting to new demands, new technologies and new structures. However, since that time the pace of change has speeded up. Many of the questions they raised about the future of the system remain unanswered while new challenges have emerged. Although fresh apples do not face the problems of extreme perishability and short seasons that some produce items face, they share many of the other difficult challenges to their marketing system.

ROLE OF MARKET INFORMATION

One challenge is the continuing imbalance in market information at different levels of the marketing system. Government continues to provide detailed price information on a same-day basis on many varieties, grades and sizes of apples at the shipping point and wholesale levels. In contrast, only a single retail price for sample cities is available on a monthly basis with several months' lag. However, it is not uncommon for a single retail chain to account for 50 percent or more of grocery sales in major metropolitan markets. From their own Universal Product Code scanning or other internal computerized sales records, these grocery chains can generate a complete information system on product flows. They are able to experimentally test various pricing and merchandising strategies and pursue alternative goals such as gaining market share, maintaining a low-price or a premium-price image, or increasing the contribution of a particular item to gross profits. Wholesalers can use their special relationship with individual retailers to gain similar data, but shippers, in general, can only get such information through expensive, continuous surveys.

The dispersed population of small shippers has usually been unwilling to pool their resources to buy such data because they cannot see how they as individuals can exploit such data. Unaware of the problems resulting from the superior information available to buyers and not to sellers, they have not lobbied government to provide such information. Retailers, naturally, oppose governments attempting to compile such information, knowing full well that it would reduce their advantage in negotiations with sellers. Governments, in turn, under budget pressures, are reluctant to institute expensive new data collection systems for the multitude of produce items now available.

Two groups at shipping point have the resources and the perspective to see the benefit from improved information further along the marketing channel. Agencies such as the Washington Apple Commission or the New Zealand Apple and Pear Board have increasingly funded market studies either to aid in targeting their promotional dollars or in sending back signals to their affiliated growers on changing market preferences. Likewise, large multicommodity organizations, which have a constant stream of produce items moving to the consumer and the ability to implement needed reactions to market preferences, have better resources to fund information gathering and are more likely to be able to earn a payoff from the information gathered. The bigger the share of produce items that these multicommodity organizations control, the less likely they are to want to share in industry information gathering efforts which provide information equally to large and small firms. On the other hand, they are more likely to be able to afford their own comprehensive information collection system and to want to use it solely for their own competitive advantage.

Improved market information throughout the system could also reduce the major problem of produce pricing: that prices at different levels of the marketing channel frequently give contradictory price signals. Usually, this occurs when rising retail prices are signalling consumers to reduce purchases at the same time that falling shipping point prices are signalling increased supplies.

Handy and Padberg (1971) argued that in the case of food processors a dichotomy was arising between an oligopolistic core with a few large food processors and a fringe group of numerous small processors. The large processors tended to trade with the large multiunit retailers while the small processors focussed mostly on serving the retail fringe of numerous small retailers. The same sort of dichotomy appears to be arising among shippers of fresh produce. The movement is being hastened along by the growing importance of electronic data interchange in national and international commerce.

In the traditional fresh produce transaction, the buyer telephoned an inquiry about price and availability to one or more shippers. At some point in the process, the buyer found a deal that was acceptable and placed an actual order. The details of the order might include quantity, quality, price, shipping instructions, terms of payment, type of transportation and who pays the transportation agency. These details became part of the written record of the transaction and were included in the commercial invoice mailed to the buyer. These details were reentered into the sellers' books as accounts receivable and the buyers' books as accounts payable. In due course, a check would be mailed by buyer to seller for the amount owed, and books of both buyer and seller would be balanced. Changes in inventory would be estimated from an analysis of recorded sales or purchases. Where sales, purchasing or accounting records were computerized, the same data would be entered many times for different analyses or accounting purposes.

Electronic data interchange recognizes that the majority of each computer's output is another computer's input, and that much effort is wasted on data entry and reentry. Under electronic data interchange, the buyer's computer can talk directly to the seller's computer and to that of the transportation agency. An order originating with the buyer can be instantaneously transferred to the seller's computer and the shipping schedule to the transportation agency's computer. The buyer's inventory can be automatically increased by the size of the order, and accounts payable increased by the value of the order. The buyer's computer can be programmed to write the appropriate check upon notice of satisfactory delivery or to transfer the funds electronically to the seller's account. The same order can automatically trigger billing in the seller's accounts department and instructions to the shipping department. Upon notice of shipment, inventory can be automatically adjusted, and plans laid for future withdrawals from storage, additional packing, and future materials supplies. Upon receipt of payment, the seller's account can be automatically adjusted.

Electronic data interchange can streamline the order process and give buyers and sellers complete information on the location of inventory. It can also permit each member of the system to automatically analyze the performance of its trading partners relative to competing firms. For example, the on-time performances and rate of damage incurred by individual trucking companies can be analyzed. However, electronic data interchange will also allow firms to become more selective in the partners with whom they do business.

There may be a tendency for large entities with more sophisticated and compatible computer systems to prefer to trade with each other. This will give a considerable advantage to the large independent shippers of fresh apples, to large multiproduce shippers who already have systemwide computing facilities, and to centralized marketing agencies such as the New Zealand Apple and Pear Board which already rely heavily on an integrated information system. The computer will be used not just to keep track of orders, sales or accounts but to identify those orders that are least profitable. For example, the cost per transaction will tend to be greatest for small or unique orders, for emergency orders or for orders with short lead times. Larger companies will either find themselves avoiding such orders or hiring off such activities to other entities who can serve those special niches.

Pressure will continue from buyers to minimize warehouse space and inventory size at point of sale and for just-in-time deliveries. Here again, new technologies in satellite tracking will make it possible to know at any time where a truckload or railcar of produce is located and to constantly upgrade expected time of arrival. Improved information on logistics will become a competitive tool among shipping agencies for the business of shippers and among shippers for the business of buyers.

Chapter 7

Retailing

The retail store provides the final test of a successful apple marketing program. It is only after the consumer buys apples, takes them home, consumes them, is satisfied, and returns for more, that the task begun many months earlier by the grower in a distant orchard is finally completed. It is at the retail level that the shopper faces a myriad of choices: to buy a fresh fruit or an alternative snack, to buy fresh apples or an alternative fresh fruit, or to buy one kind of fresh apple or another.

While a number of developments since the 1970s have favored the overall growth in demand for produce, competition in the retail battleground for fresh apples has become more intense and is likely to continue to intensify. In general, per capita and household incomes have risen in major apple consuming countries in recent decades. Numerous studies have shown that as incomes rise, consumers increase the number of produce items they buy, and those who buy any produce item, such as apples, tend to buy a wider selection of apple varieties. Thus, while the general environment for total produce sales is favorable, the challenge to apples from other produce items is growing and the competition between apple varieties is increasing.

In this situation, it is becoming even more critical for apple growers to understand what consumers are seeking and to make sure that their product is appropriately handled throughout the marketing chain and onto the retail store shelf. Then, growers or their agents must work with the retail store to ensure that their products are presented, handled, priced and promoted in such a way that consumers come back repeatedly for more. Too often in the past, growers and shippers have shown little concern for or understanding of consumer or retailer needs. In the future, however, a close understanding of consumers and good relationships with retailers will separate the successful from the unsuccessful growers.

Understanding between growers and retailers does not come easily. Growers are usually small, independent businesses specializing in one or two products. Retailers frequently form large, multiunit businesses handling thousands of products. Growers incur large costs in producing apples before they have any guarantee that they will receive sufficient revenue from sales to cover those costs. In contrast, retailers, like shippers, wholesalers, brokers and transportation agencies, have the option of refusing to handle apples unless the likely revenue covers expected costs. Growers are residual price takers, that is, the charges for retailers, wholesalers, etc., are deducted from the consumer price and the grower gets what is left. In some cases, the grower may actually be presented with a bill when the residual price is insufficient to cover packing charges. Conversely, the grower bears the responsibility for any dissatisfaction with the product along the marketing chain. For example, if the retailer complains about the delivered quality of a product and refuses to pay or seeks a price reduction, that price adjustment will usually be passed back to the grower.

At the grower level, supply is determined by the quantity harvested or in storage. Price tends to respond to the quantity placed on the market in any period. In contrast, retailers generally set fixed prices for every item in the store, including apples. Consumers react to these prices by adjusting the volume of product they buy. Retailers can use fresh apples either as a discount item to stimulate store traffic or as a premium-priced item to increase gross margins. These decisions may be tied to overall company strategy rather than to the current supply conditions for fresh apples.

TYPES OF FOOD RETAILERS

Not all retailers are the same. Fresh apples are still sold in conventional supermarkets, in giant discount warehouses, in specialty produce stores, and from wheelbarrows on city street corners. Alternative retailer outlets are continually evolving. The structure of the retail trade and its practices at any time reflect major social forces and lifestyle trends. Customers are not solely interested in the price or availability of apples or carrots or canned beans, but in the availability of parking, the speed of the checkouts, the opportunities for socializing or other features that the retail outlet provides.

The United States has been the leader in adapting its retail food system to changing customer lifestyles. However, many innovations from North America have been adopted and adapted to local conditions in Europe, Latin America, Australia, the Middle East and Asia. Certain major trends can be identified, although some retail formats remain resistant to newer trends while others become reincarnated in modified guises.

In the 1860s, retail food stores were operated by individual proprietors and their families. These were small businesses just like individual farm operations. Products such as fresh apples were hauled to local assembly points, delivered by retail to wholesale distribution centers and from there hauled by horse-drawn wagon or cart to local wholesalers or retailers. Most shopping was done daily on foot at the neighborhood store by stay-at-home housewives. Only farmers or the affluent had access to horses and the means to buy in bulk. The retail store was a social meeting place for the community. It was often a source of credit from payday to payday, or from harvest to harvest. The system worked because the shopkeepers knew their customers and vice versa.

Although the system may sound idyllic in retrospect, customers found many reasons for dissatisfaction. The standards of store hygiene and quality of product left much to be desired. Customers then, as now, were generally suspicious of the size of margins enjoyed by the transportation, wholesaling, and retailing sectors. In addition, credit was a two-edged sword. While it helped families avoid hunger in times of financial stress, it could be used to hold customers captive. Interest rates were felt to be exorbitant. In turn, many small retailers found themselves in thrall to the growing market power of major wholesalers at the major railroad terminals.

During the second half of the nineteenth century, a number of retail innovations were developed to meet these concerns. One offshoot of the cooperative movement in Britain was nonprofit retail food cooperatives. For a small membership fee and contribution of work time, members could select their own board of directors and manager and share in year-end dividends. Retail food cooperatives flourished where there was the biggest actual or perceived overcharging by independent retailers. Branch cooperatives were able to draw on the joint buying power of cooperative wholesalers and to get advice on store design and layout, pricing and merchandising. As urban populations grew and society became more depersonalized, private investors saw the opportunity for establishing chains of retail food stores with uniform, high standards of facilities, product, price and service. Organizations such as Liptons in Britain and A & P in the United States became food mass merchandisers who were no longer overpowered by the large terminal wholesalers.

Many independent retailers succumbed to the competition from chain retailers. This created a legacy of antagonism between chains and independents which persisted for decades. The chains, in turn, found themselves threatened by the emergence of the supermarket in the 1920s. Until then, the size of retail outlets was small, the number of items limited, and shoppers had to be waited on individually for each item. However, the growing affluence of shoppers, the growing number of branded items available, and the rising cost of labor led innovative retailers to experiment with various forms of self-service.

The supermarket capitalized on the self-service concept, the wider ownership of automobiles and the availability of cheap suburban land to sell a wide variety of items at significant price reductions. Conventional retailers, by attempting to hold price levels up through existing laws on retail price maintenance, offered the supermarket discounters a huge competitive opportunity. The combination of self-service, supermarket and chain concepts became the basis for the major retail food enterprises of the twentieth century.

By the mid-1950s both small retailers and legislatures throughout the United States had abandoned attempts to enforce retail price maintenance (Hiemstra and DeLoach, 1962). Smaller retailers fought back by joining cooperative or voluntary buying groups. These groups usually centered around one or more large independent wholesalers who assisted retailer members in revamping their operations into a supermarket format. By 1960, supermarkets, whether chain or independent, accounted for only 10 percent of U.S. retail food stores but for 70 percent of retail food sales. They were the dominant outlet for fresh apples and other apple products. During this period, small retailers continued to press for antitrust enforcement to curb increasing concentration in the supermarket business.

The supermarket revolution had been based on low prices through cost control and economies of scale. The early supermarkets had few frills in the form of additional consumer services. After World War II, competition between supermarket chains increasingly focussed on services such as parking, air conditioning, refrigerated shelving, and check cashing. Since then, a number of other entities have picked up the mantle of price discounter under various formats and names such as discount stores, warehouse stores, price clubs, etc. Supermarkets have responded by building combination stores and hypermarkets carrying both foods and nonfoods.

Another variant in the 1960s was the introduction of convenience store chains, such as 7-11 and Circle K. These franchised outlets, which enjoyed chain buying power, carried only a limited selection of items, remained open for long hours and charged a premium price for this service.

The increase in consumer preference for fresh produce and the concern about natural foods and fear of food additives in the 1970s led to an upsurge of demand for fresh produce at the retail level. Supermarkets responded both by adding produce space and by building bigger stores. Convenience stores and most discounters reacted by ceding most of this market to the supermarkets and by emphasizing nonfood sales. There was a surge in direct marketing in the 1970s to exploit the growing preference for fresh produce. The most important outlets were on-farm pick-your-own, roadside stands and farmer's markets where farmers and consumers could meet for direct exchange. Growth was limited because there were only a few parts of the country in which there were enough farmers in close enough proximity to consumers to make direct marketing viable (O'Rourke, 1980).

Another major factor in food retailing has been the growth of the away-from-home food market. The increase in the number of women working, the long commutes, increasing travel for work and leisure, and increasing availability of food at almost any spot where groups collect, have all contributed to the growing share of expenditure on food away from home. The major categories include schools and colleges, hospitals and nursing homes, factory cafeterias, airlines and other locations with relatively captive audiences.

However, the most dramatic changes have taken place in the commercial food service segment with the growth of what has become known as "fast food" outlets. The dominant firm, McDonald's, is typical of the sector. It offers a limited menu at competitive prices. Through franchising, it is able to have outlets at many convenient locations. Heavy national advertising is used to remind consumers of the convenience, quality, fast service and low price. Fast food outlets have been able to respond rapidly to changes in population distribution, food preferences, time limitations and labor availability.

By 1990, while retail food store sales in the United States exceeded $367 billion, sales through commercial eating and drinking places exceeded $182 million (U.S. Department of Commerce, 1991). Limited menu operators, primarily fast food operations, accounted for over 42 percent of sales by commercial eating and drinking places (NRA, 1990).

By 1990, too, the clear-cut division between different kinds of food outlets had broken down. Grocery stores were competing with fast food outlets for take-home food business, and were remaining open longer to compete with convenience stores. Gasoline outlets were competing with convenience stores. Drug stores and variety stores were selling foods. Taverns were providing meals.

In general, however, sellers of fresh produce (including fresh apples) have not been very successful in making their product available in the many new retail settings. Consistently, U.S. consumers continue to buy most of their fresh apples from retail supermarkets. Of the respondents interviewed in the 1987 Fresh Trend Survey, 82 percent said they buy most of their produce at conventional stores (Zind, 1987).

PRODUCE PRICING AT RETAIL

Supermarkets continue to be the major interpreters of what consumers want in a fresh apple. They decide how many varieties will be stocked, what grades will be supplied, whether the product offerings will be bulk or packed, and whether brand or origin will be stressed or ignored. Each week, the fate of fresh apples hinges on the success of apple marketers in the battle for shelf space, in gaining in-store promotional displays and in competing for space in weekly newspaper advertising. Supermarkets will continue to be the major link between apple suppliers and apple consumers. Accordingly, it is critical to understand how supermarkets operate and how the outcomes for apple marketers can be improved.

Once again it is important to note that even the smallest supermarket carries a minimum of 10,000 separate items, including 200 produce items. The average supermarket will carry close to twice that number and the largest supermarkets many times that number. Most supermarkets are affiliated to chains which do centralized buying, pricing and merchandising. It would be a logistical nightmare to adjust retail price each time there was a change in any of the many shipping point prices for produce. Thus, retailers must adopt longer-term pricing strategies, rules of thumb, or what McLaughlin and Pierson (1983) call "standard operating procedures." These strategies must be flexible enough to deal with competing retailers, changing seasons, the influence of holidays and weekends, weather disruptions and unforeseen speeding up or slowing down of sales of specific items.

In a classic study of retail pricing, Preston (1963) discussed the complexity of retail pricing and the difficulty in explaining much of the variation observed. Rarely would it be profitable for a retail firm to charge a constant percentage markup because of its ability to "increase sales of many products by offering attractive prices for a few." A store need not compete directly on price with its competitors if it can develop a price structure and product assortment that enable it to maintain its place in the market without uniformly matching the prices of its rivals. A group of food stores, or a chain which sells both foods and nonfoods, clearly will have considerable discretion in how it prices any individual item, as long as the pricing of the other thousands of items is sufficient to cover costs and assure a normal profit.

Hiemstra and DeLoach (1962) argued that pricing policy could not be viewed independently of overall company policy. They cite the five objectives of pricing policy identified by Kaplan, Dirlam and Lanzilotti (1958) for all business as: (1) achieving a target return on investment, (2) stabilizing prices and margins, (3) maintaining or improving market position, (4) meeting or following competition, and (5) achieving product differentiation. Which objectives are being pursued most consistently at any time by food chains will reflect current or expected conditions. For example, new entrants may be attracted by the potential return on investment relative to other possible investments. Once in the market they may have a target market share which they wish to achieve to run their facilities at optimal capacity. They may attempt to differentiate their chain from existing firms by providing special services, unusual decor or discount pricing. In turn, existing firms may seek to beat back competition or to stabilize prices and margins as rapidly as possible. Apple marketers who must sell to retailers in many countries can expect to meet varied combinations of these objectives in any single season.

In the United States, certain patterns of price and margin setting have evolved in the supermarket system. For example, a case study reported in the July 1975 issue of the *Progressive Grocer* found that produce items accounted for 5.1 percent of weekly sales, but 8.6 percent of weekly margin. This resulted from the higher percentage gross margin in produce (34.9 percent) than in all items (20.9 percent). Of the store departments studied, only general merchandise had a higher gross margin (38 percent) than produce. The meats, dairy, ice cream and grocery departments had margins less than 21 percent. Wine and frozen foods departments had margins just above 25 percent. Health and beauty aids had an average gross margin of 28.6 percent. Clearly, the higher margins on produce items made an important contribution to store profitability. A 1981 study reported in the *Produce Marketing Almanac* (1981) found gross margins on fresh apples of 35 percent, compared to 27.5 percent for bananas, 37.5 percent for oranges, potatoes and tomatoes and 32.5 percent for lettuce.

Growers of apples and other produce items consistently voice a number of complaints about the retail food distribution system. One is that retail prices do not accurately reflect prices at shipping point so consumers get misleading signals about available supplies. Another is that retail margins between purchase prices and selling prices are excessive. A third is that poor handling of fresh apples in retail distribution centers, store backrooms, and display stands causes quality deterioration, diminished value, and reduced revenue which eventually affects growers' returns. While growers tend to attribute these results to the bargaining power of individual chains, many economists have suggested that various forms of collusion among chains lead to price or margin excesses.

Unfortunately, the data needed to analyze such charges is rarely available. It would seem to be a very simple problem to find out the price paid for an apple to the grower, the shipper, the wholesaler or the retailer and to subtract the difference. However, some allowance must be made for losses due to discards, theft, deteriorating quality and shrinkage in the market system. Growers deliver in bins, shippers sell packed boxes, retailers sell single apples by count or weight. A standard bin can contain between 750 and 1000 pounds of fruit. A standard packed box may contain between 42 and 47 pounds of fruit. Even if each measures prices by weight (pound or kilogram), the price of apples at shipping point varies by variety, grade and size. A typical packer will quote a different price for 12 or more items of each variety. For example, Extra Fancy Red Delicious size 80 will sell at a different price from Fancy Red Delicious of the same size. However, few individual retailers will stock more than two Red Delicious selections. Apples close in size may be included in a single display. Extra Fancy apples may not be distinguished from Fancy. Indeed, a number of different varieties of apples may be sold at identical retail prices even though the free-on-board (FOB) price of each is different.

Even when the price of individual items can be traced, the result will vary from week to week or from chain to chain. Price specials may be offered in every outlet of a chain, in only some cities or in only some stores as manager's in-store specials. Economists have devised methods for comparing typical or average relationships between prices at different levels of the marketing chain.

Average prices for all apples, or for particular varieties, are measured at each level of the marketing chain on a weekly, monthly or other periodic basis. Ideally, the price of all apples at each level would be determined. However, from a practical standpoint, price data can be collected only for a sample of all apples. For a valid comparison between different levels or between different weeks, an identical mix of grades and sizes must be included in each sample. The same must hold true for a valid comparison of prices between cities. In addition, if one wants to compare price at any shipping point with the average retail price for a whole country, city prices must be given the weight in the calculation appropriate to their share of national sales.

A further major problem arises due to the passage of time as apples move through the marketing system. For example, lots of apples included in a pricing sample at shipping point in one week may not appear on most retail shelves until one or more weeks later. Part of a lot may be on advertised special while part may be sold at everyday prices. If all apples were sold at retail exactly one week after being priced at shipping point, it would be reasonable to compare shipping point price in any week with retail price the following week. Studies by Edman (1964), Burns and Edman (1970) and Burns and Podany (1975) used such a methodology. However, the real world is never so simple. Sophisticated analytical methods can be used to test the relationship between price series, but few in the produce industry can employ such techniques on a regular basis.

PRICES AND MARKETING MARGINS

Despite all these caveats, the United States Department of Agriculture in the 1930s pioneered a number of comparative price series for apples and many other food items which give a broad-brush picture of price relationships and marketing margins.

The first series of price spreads for fresh apples were published in 1936 and subsequently quarterly and annually for many years thereafter. For example, a 1957 USDA report on Farm-Retail Spreads for Food Products analyzed price spreads for apples for the years 1935-56. Over the period, retail prices rose from 5.7 cents to 15.1 cents per pound, farm value rose from 2.1 cents to 6.3 cents for the 1.08 pounds at the farm level needed to generate 1 pound at the retail level. The farm-retail price spread rose from 3.6 cents to 8.8 cents. However, the farmer's share of retail price rose only from 37 percent to 42 percent. During the two decades, the farmer's share varied from a low of 37 percent in 1935 and 1938 to a high of 55 percent in 1944 and 1946.

Seasonal data were reported for the years 1947 to 1957. In general, both retail prices and farm values for each crop year were lowest immediately after harvest in the October-December quarter and highest in the April-June quarter as apples became more scarce. The farmer's share of retail price was highest in either October-December or January-March when farm price was lowest. Not for the last time it was apparent to apple growers that a favorable farmer's share was not necessarily associated with higher farm prices.

Essentially the same series was continued until June 1978, when the retail price collection was discontinued for economic reasons. In the last full calendar year, 1977, retail price had risen to 39 cents per pound, farm value to 12.8 cents and farm-retail spread to 26.2 cents. Farm value approximately doubled in 20 years but farm-retail spread tripled. Farmer's share, which exceeded 40 percent in every year between 1941 and 1957 never once exceeded 40 percent in the next two decades and had fallen to 32 percent by 1977. Thus, during the period when supermarkets were expanding their dominance of food distribution and extending the services offered to consumers, the farmer's share of fresh apple retail prices was trending downwards. In each of the 20 complete crop years, 1957-77, retail price and farmer's share was lowest in the October-December quarter. In 18 of the 20 years, retail price and farmer's share was highest in the third quarter of the year, July-September. About half the time farm value was lowest in the October-December quarter, but in other years it reached a low point in later quarters.

Ironically, farmer's share was frequently highest in the October-December quarter when absolute farm price was low, and lowest in the July-September quarter when farm price was high. The within-season variation of farmer's share was quite wide, varying on average by about 10 percentage points. In 1964, farmer's share varied from 42 percent in the second quarter to 24 percent in the third quarter. In 1966, farmer's share varied from 43 percent to 25 percent. Clearly, these wide variations could not be explained solely by the influence of supermarkets on the food distribution system.

For the years 1966-77, farmer's share of retail price for apples averaged 33.4 percent, about the same as that for major vegetables such as cabbage, carrots, celery, cucumbers, lettuce, onions, peppers, potatoes and tomatoes. Farmer's share for grapefruit, oranges and lemons was 23.1, 24.1 and 27.2 percent, respectively. The farmer's share for grain products and canned vegetables was generally 15 percent or less.

MARKETING MARGINS BY VARIETY

While aggregate data such as that discussed above are useful for examining long-term trends, it is not applicable to any specific variety or point of origin. The actual margin behavior can be more easily determined for a specific variety. Because of its prominence in the United States fresh apple market, most available marketing studies have focussed on Washington Red Delicious apples. For example, Edman (1964) and Burns and Edman (1970) examined the prices and spreads for leading varieties of apples, grapefruit, grapes, lemons and oranges sold fresh in selected markets in the 1950s and 1960s. Burns and Edman found that on average the return to grower and packer (equivalent to shipping point price) of Washington Red Delicious apples was 44 percent of the average retail value for the 1962-67 period. The shipping point-wholesale spread was 16 percent and the wholesale-retail spread was 40 percent. Spreads were quite similar for other fresh fruits.

In a follow-up study, Burns and Podany (1975) examined marketing margins for a selection of apple varieties and origins and for other fresh fruits. They compared Washington Red Delicious, Eastern Red Delicious, Midwestern Jonathan, and Washington Winesap apples. Results for Washington Red Delicious were similar to those for the previous period studied. The wholesale-retail spread averaged 43 percent, transportation charges 10 percent and grower-packer returns 47 percent to Chicago and New York. Retail prices of Eastern Red Delicious averaged 20 percent less than those of Western Red Delicious. The grower-packer share (56 percent) was considerably higher but the shipping point-wholesale spread (seven percent) and wholesale-retail spread (37 percent) were lower. Midwestern Jonathan price was even lower. Since the major market was in nearby Chicago, the shipping-point wholesale spread was also seven percent but the wholesale-retail spread averaged 43 percent. Washington Winesaps, which were generally sold later in the marketing year, received retail prices about equal to those of Red Delicious but the shipping point-wholesale spread averaged 17 percent and the wholesale-retail spread averaged 34 percent. Since the cost of transportation would be the same for Washington Red Delicious and Washington Winesaps to any destination, clearly wholesalers were charging more to handle the more specialized Winesaps.

As in previous studies, it was clear that transportation costs, wholesale markups and retail margins were more stable than either retail or packer prices. Transportation charges tended to increase in line with inflation. Wholesale-retail margin also appeared to have a strong inflationary element. Grower-packer return, being a residual after marketing charges were paid, frequently did not compensate for inflation. However, the wholesale-retail margin did show considerable variation both absolutely and as a percentage of retail price. For example, for Washington Red Delicious delivered to New York City, the wholesale-retail margin varied from 38 percent in the 1967 crop year to 53 percent in the 1969 crop year. In general, the bigger the crop, the lower the shipping point price and the higher the percentage wholesale-retail margin.

A new series of data in the 1980s made it possible to analyze marketing margins for Washington Red Delicious in selected northeastern and northern central cities (USDA, August 1991). On average, for the 11 seasons 1980 through 1990, the shipping-point wholesale spread was 18.7 percent of retail price to northeastern cities and 16.1 percent of retail price to northern central cities. The wholesale-retail spread was 37.4 percent in the northeast and 39.5 percent in northern central cities. The grower-packer share of retail price was 43.9 percent for apples shipped to northeastern cities and 44.4 percent for those shipped to northern central cities. This was not markedly different from the grower-packer share in the late 1960s and early 1970s.

Analysis of monthly prices of retail, wholesale, and shipping point indicated that prices at all levels tended to move approximately in synch. Prices each season tended to peak in mid-summer and to reach a low point in mid-winter. Marketing margins, however, tended to be relatively constant. They were most likely to be out of synch at the beginning or the end of the crop year or when there was a sudden change in shipping point price.

These data on marketing margins do not definitively answer the question raised earlier about whether retail margins are "excessive" or retail prices reflect shipping point price "accurately." Margins on produce are higher than those on most other items, and have stayed consistently at high levels over time. Retail prices do tend to move in synch with shipping point and wholesale prices most of the time. The gap between retail and shipping point prices, which had expanded rapidly when the supermarket industry was increasing its dominance of the retail food trade, became more stable in the 1970s and 1980s as supermarkets faced increasing competition from alternative outlets.

While the mix of organizations selling produce has changed over time, the operations of each type of organization have also changed. For example, supermarkets have continued to offer new services, both in response to legislative pressures and to compete for customer loyalty. For example, Kaufman and Handy (1989) listed 15 services which were offered by at least 25 percent of supermarkets surveyed, and 10 which were offered by over 60 percent of supermarkets surveyed. Some of these, such as coupon redemption, bagging and check cashing, had been offered widely for decades. Others, such as bottle deposit, unit pricing, full-service delis, and scanning were of more recent vintage.

Even within the produce department, there have been significant changes in supermarket operations. As the average size of stores has increased, the average shelf space devoted to produce has grown. The share of that shelf space that is temperature-controlled has also grown, and the sophistication of the control systems has improved. These advances all increase the fixed costs per store. They can only be offset by increased volume per store that lowers unit costs. Inevitably, as more stores in a market introduce these amenities, it becomes difficult for all to increase volume per store.

The biggest single cost item in supermarket operations has been labor. For example, England (1959) reported that costs of pay and benefits exceeded 50 percent of total grocery chain expenses in each of the years 1955 through 1958. Since then, grocery chains have worked assiduously to reduce labor costs by reducing personnel, by multitasking and by the hiring of more part-timers. However, continuing increases in wage rates, in state and federally mandated fringe benefits, and in hours of operation have conspired to keep labor costs as the major expense of retail chains. In a long running series called "Operating Results of Food Chains," it was reported that while gross margins consistently averaged about 22 percent of sales, total payroll expense rose from 10.14 percent in 1962 to 13.12 percent in 1982 (German and Hawkes, 1983). Employee benefits were the major contributing factor, rising from 1.38 percent of sales to 3.22 percent. Of total payroll expenses, 75 percent were incurred in store operations.

The produce section is at least as labor intensive as other major departments. For example, certain products such as lettuce, leafy vegetables or herbs can deteriorate rapidly with customer handling, and need sorting or trimming. In an Economic Research Service study it was reported that while labor accounted for only 37.9 percent of retail expenses for fresh potatoes and 49 percent for oranges, it accounted for 75 percent of retail expenses for lettuce (USDA, 1977a).

The rapid increase in energy costs during the 1970s affected the cost of retailing in a number of ways. It increased the costs of transportation, heating and refrigeration, and packaging materials derived from petroleum products. The rate of increase slowed dramatically in the 1980s (Dunham, 1990). While the total marketing cost index rose 34.5 percent in the 1980s, all fuel and power costs rose only 10 percent. The index of labor costs in retailing rose by only 15.7 percent. Price indexes rose over 50 percent for paper bags and sacks, advertising, communications, water and sewage, business services, and property taxes and insurance. Some or all of these added costs may be passed on to the consumer.

There is rarely sufficient data available to determine whether or not increased costs are passed on to customers or are reflected in lower prices at shipping point. Due to the efforts of Richard Bartram, who was then an extension agent in the heart of apple country in Wenatchee, Washington, quality and price data were obtained for Red Delicious and Golden Delicious from regular and controlled atmosphere storage, and for Fancy and Extra Fancy grades in Washington state and in five major retail markets: Chicago, Baltimore,

New York, Los Angeles, and Houston. Trained observers selected samples from the back room of each retail store and measured weight, number and size of bruises, pressure, soluble solids, acidity and price per pound (Al Saffy and O'Rourke, 1984).

Little consistency was found between marketing margins for the different varieties, storage types, grades or sizes. However, percentage marketing margins were higher in April than in February, and for fancy grade and regular storage apples, which are usually of lower quality. The data permitted limited tests of a number of economic hypotheses that might explain the short-term differences in marketing margins, namely: (1) differences in marketing costs, (2) quality differences, (3) economies of scale, and (4) exercise of market power by retailers. The only hypothesis that could not be rejected was that in the short term, retailers could exercise market power to the extent that they did not very effectively or rapidly pass on price changes at shipping point to their customers.

Part of this power may arise from an imbalance of information. Shipping point price information is available daily on the price and shipments of different varieties, grades, storage types, sizes and pack types. However, no comparable data are available on changes in price or consumer movement in major retail markets. A retail chain with several hundred outlets in a number of major markets has access to significant retail sales data to which shippers are not privy. Other evidence suggests that retailers do not exploit their information advantage fully because they lack the personnel to analyze the millions of bits of information generated in a chain each day. In the future, as the exploitation of scanner data becomes more sophisticated, chains will acquire that analytical capability. In the meantime, however, their retail pricing is not geared to reflect each nuance in price and availability of each commodity, but to ensure a reasonable cushion of profitability under most variations that are likely to arise in a specific season.

RETAILER PERFORMANCE

A number of economists and other analysts take a more jaundiced view of the role of food retailers, supermarket chains in particular, in the food business. Of greatest concern has been the increasing concentration of retail food sales in the hands of a small number of retail food chains since the advent of the supermarket in the 1920s (National Commission on Food Marketing, 1966). For example, Parker (1976) reported that in the largest 21 standard metropolitan statistical areas in the United States, the share of grocery sales accounted for by the four largest chains went from a weighted average of 44 percent in 1954 to 48 percent in 1972. The level of concentration increased in 16 of 21 major markets. Parker and others argued that this increased concentration led to higher prices, higher gross margins and higher profits than would otherwise obtain.

The difficult analytical issue is with what to compare the current situation. Economists frequently make estimates of what would obtain if perfect competition prevailed. In such a world, all retail firms would be small, none would exercise market power, there would be free entry and exit of competing firms, and all would have perfect information. However, it is not possible to experimentally test how such a system would work. One could not afford to halt all current food retailing in a major market while one tested a system closer to the "perfect" model. So analysts attempt to measure deviations from the ideal system indirectly.

Marion et al. (1977) analyzed a unique set of data on retail food chains collected by the Joint Economic Committee of Congress. They concluded that the net profits and prices of large food chains were positively and significantly related to market concentration and a chain's relative market share. They argued that higher profits were not due to efficiency and lower costs. Rather, in more concentrated markets, retailers had higher costs because of inefficiencies and cost-increasing forms of competition such as advertising. Their results were criticized primarily for drawing sweeping conclusions from limited data.

More recently, Kaufman and Handy (1989) examined data derived from a nationwide survey of supermarket prices and other store, firm and city characteristics. They found that

> Supermarket size and sales volume, occupancy costs, store services, and (presence of) warehouse stores contributed to

firm price differences within cities. Market growth, market rivalry, and market entry accounted for firm price differences between cities. There was no evidence that firm market power—the ability to unilaterally raise prices–had a significant effect on supermarket prices.

Clearly, such indirect measures of the effect of market power on food prices can be influenced by the type of data collected and the survey coverage. While it may not be possible to collect enough data to give effective coverage over many markets and many seasons, an important question is whether or not there is real need for concern. For example, Americans continue to spend a decreasing share of per capita disposable income on food, even though the marketing services provided continue to increase. Over time, retail food prices have risen for three decades at approximately the same rate as the general cost of living. They have risen more rapidly than costs of clothing and transportation but less rapidly than housing or services such as medical care.

Within the food marketing system, while all food prices have risen three and one-half times between 1965 and 1989 (Dunham, 1990), prices for poultry and eggs rose less than three times. Meat products, dairy products, fats and oils, and processed fruit and vegetable prices rose about in line with all food. Fresh vegetable prices rose fourfold and fresh fruit prices more than fivefold. The price of food away from home has risen almost in step with the price of food at home over the last two decades. However, an individual analysis of each product line would be necessary to determine whether or not these price increases reflect demand and supply factors accurately.

On average, though, retail gross margins for food chains have remained remarkably stable over time at about 22 percent. Over the years, however, total expenses as a percent of sales have inched upwards and net operating profits downwards. Net operating profits exceeded one percent of sales in the early 1960s, declined during the 1960s, and actually became negative in the mid-1970s before recovering to over six-tenths of one percent in the early 1980s. In general, the financial strength of retail chains deteriorated after the glory days of the 1960s. The debt burden rose, the current ratio declined, other income such as cash discounts and allowances fell and net earnings after taxes fell as a percent of sales. Chains greatly reduced their expenditure on advertising and sales promotion from over two percent of sales throughout the 1960s to less than one percent in the 1980s. In addition, they increased the annual number of stockturns marginally. Thus data on retail chains appeared to reflect the increasing competitiveness of the food sector, and greater concentration on operating efficiency to meet customer needs.

During the 1980s, the retail food scene has become increasingly complicated in ownership and structure. Leveraged buyouts of some of the largest operators, including Safeway, which was then number one, have caused chains to take on onerous debt burdens, part of which were covered by selling out whole segments of the chain. Mergers and takeovers have combined large chains under a single financial umbrella, as in the case of American Stores. Often such financial restructuring has brought supermarket chains under the same ownership as rivals such as warehouse foods or convenience stores, or of nonfood retailers such as drug stores and discount stores. Much foreign capital has entered the retail food system. Of the organizations that now have strong positions in a number of retail formats, many have the financial muscle to extend their use of existing formats or add new formats. However, conglomerate ownership makes it more difficult for suppliers to evaluate the success of the different formats.

TRENDS IN FOOD RETAILING

Clearly, it is important for suppliers such as the apple industry to identify the newer trends emerging in food retailing and to adapt their marketing methods accordingly. A number of surveys have attempted to elicit where consumer wants and needs are heading and how the food retailing industry itself expects to have to adapt.

The underlying demographic changes that will occur in the 1990s are a frequent starting point for such forecasts. For example, Frumkin (1989) cites three key demographic trends identified by Ann Clurman of the market research firm of Yankelovich Clancy Shulman: the aging of the baby boomers (people born between 1946 and

1964), the continuing baby boomlet and the increase in the number of senior citizens. As they become parents themselves, baby boomers become less novelty-driven and more selective shoppers. An NRA (1988) study of 108 leaders of the food service industry estimated that household incomes would grow by 88 percent between 1985 and 2000 in households with a head aged 35 to 50. Those middle-aged households would account for over half of all growth in household incomes. Those in households headed by persons 50 and over would account for almost 40 percent. Households headed by persons under 35 would account for only 5.3 percent. Gruen (1989) pointed out that those over 55 now control most of the country's wealth, though not its income. This gradual redistribution of buying power will require retailers to adapt as much to the mobile lifestyles of baby boomers and their children as to the fading eyesight or other physical deficiencies of older shoppers. Other notable demographic trends will be the increase in minority shoppers, especially Hispanic and Asian, more single person households and more proxy shoppers.

Retailers will attempt to respond in a number of ways. Heller (1990) reported on a *Progressive Grocer* survey of 380 executives which asked them to predict the likelihood of when certain changes would occur in the retail food business. The majority expected competition to increase in the 1990s, both from other retail food formats and from eating places. They expected superstores and combination stores to increasingly replace conventional supermarkets. Scanner data would be used more widely to manage stocks and to improve returns. There would be greater pressure for logistical efficiency in ordering, delivery, and store and warehouse operations. Computerized space management and just-in-time delivery would become commonplace.

The executives expected nonfoods and food service to grow in importance while the grocery share of store sales would continue to shrink. They expected labor availability and quality to be an increasing problem. The NRA study foresaw a similar labor problem with fewer young people in total, more women in the work force and greater reliance in food retailing on minorities and immigrants with lower skill levels. Both the *Progressive Grocer* and NRA studies saw increasing pressure on their business emanating from the demands of consumers, of society and of governments. Consumers' concerns included the effect of different foods on diet, health and fitness, and of threats to health from additives, processing methods or tampering. Societal concerns included fears of energy shortages, the growing mountains of solid waste, demands for biodegradable plastic and continuing controversy over the best form of packaging. Both food store and restaurant executives expected to have to battle more government red tape by the year 2000, some related to societal issues already discussed, some related to new or increased government taxes.

All of these issues (and others which must be omitted because of lack of space) offer the apple industry new challenges and new opportunities. Clearly, any innovations which make it easier and cheaper for debt-laden chains to handle apple products will give a supplier a competitive edge. If just-in-time delivery can be offered by nearby suppliers, more distant or foreign suppliers may need to maintain intermediate warehouses from which they can rapidly fill retail orders. As larger stores stock more nonfoods, the gross margin contribution of apples will have to compare favorably with that of nonfoods. Avoiding losses due to quality deterioration and stimulating stock turns through increased promotion will become important. Product, packaging and service will have to be tailored to the different store formats. An attractive consumer pack may be needed to sell fresh apples in a convenience store, an attractive carton in a warehouse or discount store.

To service middle-aged households, taste may become more important. For older customers, easily readable labelling or convenient shelf-placement may be critical. Lighter bulk cartons may be preferred by younger workers or by women. Workers with limited knowledge of produce or limited English may need multilingual directions on proper handling methods. Suppliers will need to actively cooperate with retailers in reassuring consumers of the safety and wholesomeness of fresh apples. Ways need to be explored to find packaging which preserves apples in transit, but reduces the solid waste problem, through reduction of bulk, recycling or biodegradability.

One thing appears certain: food retailing in the United States by the year 2000 will be different in many critical ways than food retailing in 1990. The same is likely to hold true for many other countries. The self-service supermarket and many other innovations in food retailing have become well established in many developed countries and have been growing in influence in the emerging economies around the world. In the 1990s, they will spread to Eastern Europe and the Soviet Union as those economies move away from a centrally planned system. As in Hong Kong, where wet (open air) markets survive alongside modern supermarkets, each country will have its own blend of popular store formats. However, all will be affected by the consumer concerns, environmental issues, and government efforts to address them. Retailers will continue to be the sounding board for consumer concerns and the focal point for government efforts to address those concerns. It will become even more critical for apple suppliers to understand retailer needs and to adapt rapidly to meet them.

Chapter 8
Apple Processing Systems

The apple has the twin benefits of being a delightful fresh snack and a useful ingredient for many forms of cooked and preserved foods. Many of these uses have won their way into American and international folklore. "As American as apple pie" is not just a cliche, because apple pie at the same time symbolizes the best of home cooking and is the dessert of choice in many of America's most popular restaurants. Applesauce is frequently used in conjunction with pork dishes or as a staple baby food. Apple juice and apple cider are often associated with the joys of harvest and of fall celebrations. Apples are amenable to storage in dried, canned, or frozen form as flakes, chips, slices, preserves, pie fillings, and juice concentrates. Apples are also a popular ingredient in blended juice drinks, yogurt, cereals, and other products.

The apple processing system has evolved from a cottage industry which applied home recipes to seasonal supplies of raw materials to its present state, that of a technologically sophisticated, multinational enterprise drawing its supplies from all around the globe and selling its finished products in a competitive international marketplace. Its continuing transformation is shaking up many of the traditional relationships upon which the industry was based.

The era of most rapid growth in apple processing in the United States was the period from the end of World War II in 1945 to the early 1970s. An increasing number of wives and mothers were taking jobs outside the home. Family incomes were rising. Housewives were finding many more valuable uses of their time than cooking or preparing popular foods such as apple juice, applesauce or apple pies. Processors rushed to meet consumers' needs for familiar products minus the drudgery of cleaning, preparation, or cooking.

MANUFACTURING PROCESSES

The actual processes involved in manufacturing apple products are relatively simple. The quality of the product eventually produced requires selection of appropriate raw products, carefully stored and managed concentrate or other intermediate products, proper timing of heating, cooling, drying or other treatments, and blending to achieve a consistent taste and texture that will stand up in the marketing system and in normal household use.

Apple juice and apple cider are made in similar ways. The apples come from culls of fresh packing houses, from orchard run fruit or, in some cases, from drop apples. The apples are ground and pressed and filtered to remove the skins and pulp of the apple and allow the juice to flow through. Apple juice tends to be more filtered or clarified than is apple cider which has more apple particles, although apple juice is increasingly being sold in "natural" or cloudy form. Apple juice is always pasteurized whereas cider makers usually use flash pasteurization to kill any bacteria. Some cider makers use preservatives in small amounts to increase shelf life. However, all cider must be refrigerated to prevent fermentation. Apple cider can also be allowed to ferment to produce the alcoholic beverage known as "hard cider" or "applejack." Alcoholic cider is popular in the United Kingdom and Canada, but rarely used in the United States. A carbonated nonalcoholic cider drink is also popular.

The development of apple juice concentrate has given bottlers of apple juice tremendous flexibility in shipping, storage, and distribution of their product. After grinding and pressing, the essence is stripped from the raw juice which is then treated for starch and pectin removal, filtered, condensed to the desired concentration, cooled, and drained (Agriculture Canada, 1985). Some plants reintroduce the essence with the concentrate, others prefer not to. The concentrate is kept in cold storage (5° C) to prevent browning and to enable the product to travel as unrefrigerated cargo to its export destination. On average, one metric ton of raw fruit yields 20.03 U.S. gallons of 71° Brix concentrate. The concentrate is normally held in 58-gallon plastic drums for storage or shipment.

The second most important processed product, applesauce, results from the industrializing of the steps used by a homemaker in making homemade applesauce (Ho, 1974). Apples are unloaded, inspected, sorted by size and quality and by alternative uses. Decayed, crushed, or badly misshapen fruit may be placed in vinegar stock. Smaller apples may be diverted to juice. Apples for sauce are washed, peeled and cored, trimmed, chopped, and transported to the cooker where the needed sugar is added. The cooked sauce is filtered to remove seed cells and corky material and to ensure uniform particle size. It then goes to a consistometer which tests the product for consistency and Brix. Water and sugar are automatically added to bring the sauce to the desired standards for consistency and soluble solids.

The heated sauce moves from a holding tank to the can filler. The filled cans are lidded and sealed and cooled to an average temperature of 100° F. Cans may be immediately labelled, put in cases and loaded for shipment. Or, they may be stored for later labelling or shipped to other distributors for private or generic labelling. Inspection may take place after cans are closed, after they are cased, or after they are stored. Ho's study examined various factors affecting the efficiency of apple sauce processing, in particular, how computerization of information on the various steps could be used to improve efficiency and profitability.

In the intervening years, the computer has become the key to sophisticated apple processing. It is used to control sorting, accounting, sampling, inspection, inventory management, efficiency of operation and many other activities. Developments in food science and human nutrition have been applied to making processed apple products more cheaply, efficiently and in tune with consumers' changing needs.

The use of apples for processing grew steadily between World War II and the early 1970s. USDA reported data separately for apples for canned, frozen, dried, and other (including juice and cider) uses. Total apple production, which had dropped sharply after World War II, had begun a new, long-term, upward trend in 1952. Between 1952 and 1966, dried use gave way to the newer technologies involved in canning, frozen and juice processing. In the same period, canned use doubled and frozen and other (primarily juice) use tripled (Tomek, 1968).

Processors tended to locate in or near producing areas where plentiful supplies of raw materials were available at relatively low prices. They were aided by the fact that excellent processed products could be made from fruit that could not meet fresh standards because of deficiencies in shape, size, color, bruising, insect stings, surface defects or other problems. In general, raw materials were available at a price considerably below that paid by fresh marketing channels. For example, in the 1962-72 period, prices growers received for fresh apples averaged more than twice those of apples for processing (Greig and Blakeslee, 1975). Within the processing sector, generally higher prices were paid for the larger, so called "peeler" apples used in canning and freezing. Dried apples required fruit of sound internal composition and received about 80 percent of the price of apples for canning and freezing. Generally, apples for pressing into juice and cider were of lowest value, averaging about 60 percent of canning and freezing prices.

As a result, processors could manufacture products such as apple sauce and apple juice, bear the cost of storage, transportation and distribution, and deliver a product to the consumer at a price competitive with the home-made item. Technological developments made it possible to deliver other products such as apple chips or sparkling cider that could not be easily produced at home.

A number of factors in the 1970s deeply influenced the continued expansion of the apple processing industry. There were widespread concerns that in its focus on convenience, the processing industry had sacrificed the inherent taste and nutritional content of the raw product. In addition, there was a concern that processors were adding salt, sugar or other preservatives in order to lengthen shelf-life or increase product stability at the cost of food safety. Whatever the cause, in the mid-1970s, per capita consumption of almost all processed fruits and vegetables began to decline. Per capita consumption of fresh fruits and vegetables, which had been in decline for decades, began to grow. Per capita consumption of processed apples also declined, with the notable exception of processed apple juice which will be discussed later. The demand for canned apples and apple sauce was affected by the end of the baby boom and the plateau in pork consumption.

STRUCTURE OF THE APPLE PROCESSING INDUSTRY

The structure of the apple processing industry is extremely diverse and has changed over time in response to changes both in the general economy and in the world apple industry. While most apple processors began as small, local firms specializing only in apples and other local fruits, the industry now encompasses firms of many sizes and organizational forms, involved in many different fruits.

For example, Greig and Blakeslee (1975) reported that the U.S. apple juice processing industry was characterized by relatively small private or cooperative processing plants, each having local or regional distribution. Greig reported

> Some large corporate firms do process apple juice products, e.g., Duffy Mott, a subsidiary of American Tobacco Co.; Michigan Fruit Co., a subsidiary of Curtice-Burns; and subsidiaries of Pet Inc., and Borden. However, apparently only Duffy Mott and Michigan Fruit process very much apple juice. And this is primarily for regional distribution. There are few if any nationally well-known retail brands and apparently only quite limited local or regional advertising.

Processors assured access to raw material supplies by maintaining good relations with local growers. Grower-owned cooperatives had particularly favorable access to raw material supplies. A number of corporations attempted to gain a stronger foothold in the industry by conglomerate mergers with cooperatives, but with limited success.

The applesauce processing industry was also similarly tied to plentiful local supplies in the states of New York, Pennsylvania, Virginia, West Virginia, Michigan, Washington, Oregon, and California. The share of the national consumer market held by each state was strongly influenced by the cost of shipping a bulky, low value item such as applesauce over long distances (Greig, 1971).

Demand changes had a major influence on the structure of the apple processing industry after 1972, the year which marked the low point of per capita consumption of fresh apples. Thereafter, the share of apples used fresh stopped declining. Both total production of apples in the U.S. and production used fresh increased by about

60 percent between 1970-72 and 1988-90 (USDA, August 1991). However, canned use rose only by 27.5 percent, frozen use by 42.1 percent. Other uses (for vinegar, wine, jam, and fresh slices for pie making) actually declined by more than one-half. Use for juice and cider and for dried apple products rose by about 90 percent. Demand for dried apples was boosted by their use as ingredients in cereals and cake products and their popularity in diet programs. Apple juice demand was boosted by the increasing popularity of natural juices in straight juices and in blends and by the lower cost of apples for juice compared to other major fruits used for juice.

For example, Greig and Blakeslee (1975) reported that in the five seasons, 1967 to 1971, the weighted average farm price per gallon of juice equivalent was 20 cents for apples, 28 cents for California Thompson seedless grapes, 42 cents for oranges, and 77 cents for Concord and other Eastern varieties of grapes. Ten years later, in the five seasons, 1977 to 1981, the weighted average farm price per gallon of juice equivalent was 56 cents for apples, 74 cents for oranges, 111 cents for juice grapes, and 136 cents for raisin grapes (USDA, August 1991).

However, a breakthrough in the technology of concentrating apple juice for easier storage and distribution had a dramatic effect on the U.S. and world apple juice industry. The technology for concentrating orange juice had been perfected in the 1950s. It permitted orange juice to be produced in a three to one concentrate form. That is, for reconstitution, three pints of water would be added to one pint of concentrate to yield the equivalent of four pints of single strength orange juice. Frozen concentrate orange juice capitalized on the reduced transportation costs to become a nationally distributed breakfast drink. Many major food companies such as Coca Cola and Beatrice Foods managed nationally advertised brands of frozen concentrated orange juice.

During the same period, the apple juice industry was dominated by single strength juice produced by smaller firms with primarily local, or occasionally regional, distribution. Three to one frozen concentrated apple juice, despite technical improvements, was not a marketing success. However, the development of six to one concentrate, initially for orange juice, became a source of dynamic change in the apple juice industry. Overseas suppliers were first to capitalize on the cost-savings advantages in transportation and storage. Apple juice imports, which had been negligible in the early 1960s, exceeded 30 million gallons (single strength equivalent) for the first time in 1971. Much of this product was destined for use in pop wines which were then enjoying a brief period of popularity. These pop wines were made primarily from cheaper wine grapes. However, in both 1970 and 1972 there was a short crop of grapes and sharply increased grape prices. The makers of pop wine found that they could substitute apple juice imports for grapes as the basic sugar ingredient in pop wine. Greig and Blakeslee (1975) estimated that 75-85 percent of apple juice imports in 1971 and 1972 were used in the production of pop wine in California. Over 70 percent of apple juice imports entered through California ports, mostly in San Francisco. Much of the remainder entered through the port of New York, destined for New York wineries.

The surge in demand for pop wine, and the short grape crop, led to record apple juice prices in 1973, triple the levels paid to growers in 1971. In response to this record demand, apple juice concentrator capacity increased dramatically, at different stages, in countries such as Austria, West Germany, France, Switzerland, and Spain in Europe, and Argentina, Chile, New Zealand, and South Africa in the Southern Hemisphere. By 1980, an all year round supply of apple juice concentrate was available at relatively low prices. However, the primary market where this product could find a home without trade barriers was the United States. In 1981, imports of apple juice into the U.S. exceeded 100 million gallons (single strength equivalent) for the first time, and by 1985 exceeded 200 million gallons for the first time (USITC, 1986).

The growth in imports of apple juice concentrate between 1979 and 1985 radically altered the structure of the apple juice industry both in the U.S. and among its many foreign suppliers. In the U.S., the dependence of the industry on locally available supplies was shattered. Major food processors such as Beatrice Foods, Campbell's Soups, Coca Cola, Tropicana, and Gerbers were able to base nationally marketed brands of apple juice on reconstituted imported concentrate. Bottling plants could be set up near major ports of entry such as Philadelphia, New York, Seattle, Tampa, and Baltimore.

Local or regional processors or concentrators of apple juice were forced to lower their price for processed apples to affiliated growers and to utilize cheap imported concentrate to maintain the price competitiveness of their consumer packs relative to national brands. Concentrator capacity in the U.S. was either underutilized or withdrawn from production. Average price returned to U.S. growers for apples for juice fell by over 40 percent in real terms between 1977-79 and 1982-84. The average price of imported apple juice fell by 43 percent in the same period and reached its lowest level in 20 years in 1984.

The increased availability, lower raw material prices and national marketing campaigns radically altered the demand for apple juice products in the 1980s. Per capita utilization of apple juice doubled between 1975 and 1980 and increased by a further 50 percent between 1980 and 1985 (USDA, August 1991).

U.S. apple growers suffered a decline in price without any increase in volume sold as all the increase in consumer demand in the 1980s was met from imported apple juice concentrate. The share of U.S. apple production used for juice, which had risen rapidly in the 1970s, peaked at 24 percent in 1980. It exceeded that level only once during the 1980s, when there was a record total crop in 1987. The share of apple juice sold in the U.S. that derived from imported concentrate exceeded 50 percent by the mid-1980s. U.S. apple growers viewed these developments with alarm. An increase in plantings in the previous decade was expected to lead to record apple production in the U.S. in the 1990s. The apple juice sector was the only segment of demand for processed apples that was still growing. Apple growers feared that the apple juice outlet for their growing production would be lost to imports.

INTERNATIONAL TRADE COMMISSION INVESTIGATION

On December 27, 1985, the United States Trade Representative requested that the United States International Trade Commission institute an investigation of imported apple juice under Section 201 of the Trade Act of 1974. Under Section 201 regulations, if a U.S. industry could be shown to be suffering serious injury as a result of imports, it could receive appropriate relief. The American Farm Bureau Federation and farm bureaus from 28 apple producing states represented the apple industry in support of the petition. Attorneys for the petitioners argued that increased apple imports were causing actual and threatened serious injury to three domestic industries: the juice apple production of U.S. apple growers, producers of concentrate from domestic juice apples, and producers of retail processed juice products (primarily single strength juice) from U.S. juice apples.

Plaintiffs argued that imports of apples entering the United States had jumped from 43.5 million gallons single strength equivalent in 1980 to 214.4 million gallons in 1985. The market share of U.S. apple juice supplies from imports had risen from 19.3 percent in 1980-81 to 53.7 percent in 1984-85. Average price of imported juice had fallen from $1.00 per single strength equivalent gallon in 1979 to $0.64 per gallon in 1985, and had depressed the prices received by U.S. growers for apples for juice.

In addition, the flood of imported apple juice concentrate had virtually eliminated the domestic concentrating industry. Many firms had been forced out of business. Other capacity had been idled indefinitely and still other was being operated at a fraction of full capacity. Many local processors of concentrate, faced with competition from national brands who relied entirely on cheaper imports, were forced to reduce their own concentrator activities and use a greater proportion of imported concentrate in their products. Producers of retail processed juice products from U.S. juice apples were primarily affected by the generally lower retail juice prices brought about by the increased volume and lower price of imported apple juice.

Plaintiffs attempted to show that the apple juice business included three industries producing a product, apple juice, which was like the imported product under investigation. These were growers, concentrators, and reconstitutors and their employees. Damage to any one of these parts would constitute damage under the law and would merit relief. However, their argument was considerably weakened by the fact that only a minority of concentrators supported the complaint. Most concentrators preferred to retain free access to the world supply of apple juice concentrate because of the uncertainty surrounding local supplies of apples for juice from year to year. Supplies available for juicing are vulnerable both to general crop conditions and to the price prevailing in fresh and other processed markets. To maintain a widely distributed, heavily promoted brand, a continuous supply of raw materials at a stable price is desirable. Store shelf space or customer loyalty lost when product availability is limited may be almost impossible to regain. Some processors also argued that imported juice was needed for blending to maintain particular taste characteristics.

The reconstitutors of imported concentrate and the brokers who supplied them were uniformly opposed to the complaint. A successful apple juice suit could not only threaten their sources of supply, but could set an unfavorable precedent for other imported juices.

The opponents of the complaint argued that there was only one U.S. industry producing products like or directly competitive with imported apple juice concentrate. They contended that growers, concentrators, and reconstitutors represented different stages in bringing the apple juice product to the customer. In particular, concentrators and reconstitutors were inseparable, often conducting both activities in the same plant. It followed, therefore, that injury to any part of the industry would have to be offset against gains to other parts of the industry from the availability and use of imports.

Opponents argued that, in fact, injuries to the industry due to imported apple juice were negligible and were outweighed by benefits. One witness, Dr. Max Brunk, Emeritus Professor of Agricultural Economics at Cornell University, argued that while juice apples on average accounted for 23 percent of U.S. apple production, they usually accounted for only 10 percent of the value of production. He argued that growers' intentions are to raise all apples for the fresh market, and that processing-quality apples are an undesirable accident of no economic value to growers. However, this ignores a basic concept in economic theory: that of joint products. For example, in dairying, while the main output is milk, the sale of calves is a small but inseparable part of total income. In sheep production, the main output may be wool and the joint product meat. In each case, what is relevant to the producer is not whether milk generates more revenue than calves, or wool than sheep meat, but how the total revenue compares with total costs.

Exactly the same situation would appear to apply to apples. While growers do all in their power to produce apples of fresh market quality, they know that at least one quarter of production will not meet fresh market standards and will have to be sold on the lower-price processing or juice markets. However, the 10 percent of revenue earned from processing or juice apples is as valuable to the grower as any other 10 percent earned for fresh market apples. If price of juice apples falls, not just juice revenue, but total orchard revenue, falls. Since costs will not have changed, net revenue or profit will have fallen. Thus, a small percentage drop in apple juice price can cause a dramatic reduction in grower's net profit.

Other agricultural economists presented estimates of the effects of imports on prices for U.S. juice apples. Allison and Ricks (1986) argued that imports of apple juice concentrate were a direct substitute for and significantly affected the U.S. price of apple juice. This, in turn, was reflected in prices to growers for juice apples. Allison and Ricks also argued that at the margin there would be some substitution between apples for fresh and apples for processing uses, depending on relative prices. In years when juice prices are relatively high, growers will divert more low quality fruit from the fresh market to the juice market. In years when juice prices are relatively low, potential juice apples may not be harvested or dropped apples may be left on the ground. In either case, the supply of domestic apples for juice will be reduced. In a statistical analysis for the years 1971 to 1984, they found that the average grower price for juice apples in the U.S. was negatively affected by increases in U.S. apple production or in imports of apple juice concentrate. It was positively affected by increases in orange juice price or in per capita disposable income. These four factors together explained 93 percent of the variation in grower price of apple juice.

Baumes and Conway (1985) developed an econometric model of the U.S. apple market which looked at demand for fresh apples and all processed apples for the 1952-81 period. They found that utilization of all apples was higher when fresh and processed prices were higher, suggesting that "increases in the processing price are reallocating some apples that otherwise might be abandoned into the processing market." They found that the flexibility of demand for all processing apples was about one at the farm level. That is, each one percent increase in supply would depress price by one percent. These findings were consistent with those of previous studies.

To throw light on the apple juice import case, O'Rourke (1986a) adapted the Baumes and Conway model for the 1969-84 period. The O'Rourke model permitted separate analysis of the juice apple market from the other processed market and examination of the influence of imported apple juice. He found that the use of domestic U.S. apples for juice tended to be a constant share of total apple utilization and to be little affected by the grower price for juice apples. In contrast, the retail price of apple juice was most strongly influenced by the price of imported apple juice. The marketing margins for apple juice were relatively fixed. Thus, a decline in retail or wholesale price of apple juice would translate into a more severe loss to the grower per equivalent pound of apples. A ten percent decline in the price of imported juice would cause a decline of 2.25 percent in wholesale apple price and a decline of 5.80 percent in the grower price of juice apples.

Despite these economic findings, the U.S. International Trade Commission (1986) concluded "that imports increased and that the domestic industry producing apple juice experienced economic difficulties. However, we find that the domestic industry is not seriously injured or threatened with serious injury." This conclusion derived directly from the Commission's finding that there was only one domestic juice industry, not three as argued by the plaintiffs. That industry included growers, processors, concentrators, reconstitutors, and bottlers. Thus, any losses to growers or concentrators due to imports was offset by gains made by reconstitutors or bottlers. One Commissioner, David B. Rohr, disagreed with the conclusion of his colleagues. He argued that growers had suffered severe revenue losses and the concentrators had suffered major losses as they were idled while larger integrated firms or bottlers had been able to gain a competitive edge by greater use of imported apple juice concentrate.

The Commission report is a valuable source of information on the changes that had taken place in the apple juice industry. It estimated that about 25 large processors pressed 75 to 80 percent of domestic apple juice. Of these, the largest were Tree Top, Cadbury Schweppes (Duffy Mott) and Seneca. At least 16 firms had the ability to produce concentrated apple juice in the 1981-85 period. By 1985, several of these firms had closed or idled completely their capacity or were no longer in existence. At least one major company, Coca Cola, was relying completely on imported concentrate.

About half of all imports were imported directly, while the remainder were purchased through importers who rarely handled the physical product. About half of importers' sales were to companies that blended imported concentrated apple juice with domestic juice. Other major customers were reconstitutors or bottlers, producers of three to one frozen concentrate, and producers of mixed fruit juices and drinks.

The major supplying countries were West Germany, Argentina, Austria, the Netherlands, Spain and South Africa. By 1985, about 70 percent of apple juice imports entered through the New York, Seattle, Tampa, and Baltimore customs districts. The California customs districts of Los Angeles and San Francisco, which had dominated imports during the pop wine era, accounted for less than nine percent of 1985 imports.

The Commission report detailed the general increase in world apple production, the introduction of modern technology for concentration and the growing supplies of apple juice concentrate from many countries. By 1985, Argentina was processing over 40 percent of its apple production into concentrate and was exporting 97 percent of that product to the United States. Austria, Belgium, Spain, and Chile were exporting over 60 percent of their concentrate production to the United States. It was suspected that much of the exports of concentrate from Austria and West Germany originated in the then centrally planned countries of Eastern Europe and the Soviet Union. For example, Poland was reported to have become the world's fourth largest exporter of concentrated apple juice, using modern processing equipment imported from Western European sources on the basis of a barter arrangement and paid for with apple juice concentrate. Much of the increased capacity in Latin America resulted from direct investment by European or North American firms.

POST-ITC INVESTIGATION

No study of the U.S. apple juice market as detailed as the International Trade Commission Study has been conducted since 1986.

However, other data suggest that worldwide production and trade in apple juice products has continued to grow. By 1990, West Germany, Argentina, Austria, Italy, Spain, Hungary and Chile were, like the United States, still major producers of apple juice. While markets in Europe and Japan were taking a growing share of imports, the United States was still the primary target for exports of concentrated apple juice. In Europe, about half of available supply was exported and about half consumed in the country where produced. However, about 75 percent of Southern Hemisphere supply continued to be exported. In Argentina, Chile and Australia, over 90 percent continued to be exported, and over 95 percent of Argentine exports entered the United States.

While total U.S. apple production increased 17.2 percent between the 1978-80 period and the 1988-90 period, total use for processing increased only nine percent. The rate of increase varied by processing use: 9.4 percent for canned, 6.9 percent for juice and cider, 74.7 percent for frozen, and 23.3 percent for dried. However, juice and cider continued to account for about half of all processed apple use, and canned for a further one-third (USDA, 1991). Imports of apple juice, which had grown rapidly prior to the ITC investigation, grew more slowly in the second half of the 1980s and averaged over 216 million gallons single strength equivalent in the 1988-90 period, more than four times that in the 1978-80 period. Per capita consumption of apple juice, which had almost doubled between 1978-79 and 1984-85, grew only an additional nine percent by 1990-91. However, that increase was supplied primarily from increased imports.

A number of turbulent influences affected the U.S. apple juice market after the ITC investigation. The U.S. produced a record apple crop in 1987. Delays in harvesting and low fresh prices led to record diversion to processing, particularly into juice. Use of U.S. apples for juice and cider exceeded the previous record in 1980 by more than one-third, and grower prices plunged to a twelve-year low. During the 1988-89 season, a *60 Minutes* story on the CBS television network about the hazards of Alar use on apples caused a major consumer scare. Consumption of fresh apples and processed apple products, particularly apple juice and applesauce which are heavily consumed by small children, came under heavy criticism;

this despite a *Consumers' Research* (1989) report that showed Alar no more dangerous to humans than tap water. Before reason returned to the controversy and the low level of risk became apparent, consumption and prices of both fresh and juice apple products suffered a short but severe setback. Both fresh producers and processors suffered heavy losses until the inventory of Alar-treated apples was cleared.

In the 1990 and 1991 growing seasons, in both the Northern and Southern hemispheres, there was extensive loss of the apple crop among major producers. For example, in the European Community, apple production fell seven percent from 1989 to 1990, and a further 16 percent in 1991. The decrease in 1991 was particularly widespread. Similar reductions occurred in Eastern Europe and in the Southern Hemisphere. Price for processing apples began to rise in the fall of 1990 and reached record levels during the 1991 crop year.

APPLE JUICE CONSUMPTION PATTERNS

By the late 1980s, apple juice had become a major and stable proportion of the fresh juice beverage category in the United States. While orange juice usually accounted for 60 percent of the category, apple juice remained well ahead of all other fresh juices at about 20 percent (Florida Department of Citrus, 1991). Apple juice price per single strength equivalent gallon averaged 70 to 85 percent of orange juice price. While 40 percent or more of orange juice was sold in frozen form, only about 20 percent of apple juice was packed that way. Over two-thirds of apple juice was sold in ready-to-serve form.

Another analysis of juice competition noted that

> Apple juice is very strong competition for orange juice. It is perceived as having similar juice values, is used by the majority of orange juice users and is introduced into the diet at an earlier age. Ways need to be found to limit apple juice growth by making orange juice, as a category, relatively more attractive. Recent apple juice growth has occurred with relatively little marketing support; if the apple industry were to mount an

advertising effort, apple juice could become an even more formidable competitor (Florida Department of Citrus, April 1987).

Heavy orange juice users tended also to be heavy apple juice users. While 65 percent of orange juice purchasers did so once a week or more often, 44 percent of apple juice purchasers bought that often. Apple juice was more important to households with younger parents and small children. Orange juice was more heavily used by adults at breakfast only; apple juice at other occasions only.

While orange juice was more often available in commercial eating places than apple juice, apple juice was available in 80 percent or more of noncommercial eating places such as colleges and universities, hospitals and nursing homes (Florida Department of Citrus, 1988). Weakest apple juice penetration was in the largest commercial category, fast food restaurants. Apple juice availability in public school systems jumped sharply in the 1980s with incidence being highest in the larger schools (Florida Department of Citrus, August 1987). Thus, apple juice was a widely distributed consumer product with some major market segments offering considerable opportunity for growth.

Canada experienced similar developments in the apple juice market in the 1970s and 1980s (Agriculture Canada, 1985). Per capita disappearance of apple juice increased rapidly in the late 1970s and early 1980s, with an increasing proportion coming from imports. As in the United States, the proportion of domestic apple production available for processing was much higher in the eastern producing regions than in the West. However, the relatively low price of imported concentrate essentially created a ceiling on what could be paid to Canadian growers for raw juice apples. Argentina and South Africa were the largest suppliers of apple juice concentrate. The United States was the third largest supplier (of concentrate being transhipped from South America). Austria, West Germany, Hungary, and Chile were other important suppliers. Per capita consumption approximated the level achieved in the United States and was expected to increase by about two-thirds between 1985 and the year 2000.

Production and consumption of apple juice in Japan took off at a later date than in North America (Jussaume, 1991). The first apple juice-based drink marketed by a major distributor in Japan was Coca Cola's HI-C brand of a 50 percent apple juice drink which it introduced around 1972. This product competed primarily with carbonated beverages. Consumption of natural juices such as 100 percent apple juice grew in the 1980s. The share of apple juice consumed as natural juice rose from 31.5 percent in 1982 to 72.7 percent in 1989. The proportion of apple juice sold in cloudy form (as opposed to the clear, filtered form) rose from 58 to 87 percent in the same period. Many types of food companies, including agricultural cooperatives, makers of carbonated beverages, breweries, dairies, pharmaceutical firms, and others, supplied one segment or other of the apple juice market. On April 1, 1990, imports of grape, grapefruit, and apple juice into Japan were liberalized. Apple juice products would continue to be subject to duties of 25-35 percent, but could be imported into Japan by any firm or individual in any amount. Jussaume reported the import volume tripled between 1989 and 1990. The major source of imports was the United States which could supply consumer packs at very competitive prices.

OTHER PROCESSED APPLE PRODUCTS

While apple juice has become a truly international product both in bulk and final form, all other processed apple products have become less and less tied to local supplies and processing facilities and more and more dominated by the marketing needs of major food companies. These products must now meet the specific markets targeted by national food processors of consumer products, retail food chains, cereal manufacturers, generic brands, discount stores, institutional food suppliers, importers and exporters, and other emerging retail forms. Some of the hottest new branded products in the 1980s were apple based, such as sparkling apple cider, single serve packs of apple sauce, apple chips, apple juice packed in 12-ounce aluminum (Coke) cans, and aseptic juice packages which required no refrigeration. Marketing considerations rather than raw material supplies have become the dominant factors in determining the future direction of the apple processing industries.

Apple growers find themselves in the ironic situation that one of their major products, the apples which do not meet fresh market grade, but are the basis for major processing industries, rarely cover their full share of production costs. For example, in Washington State, a typical orchard might produce 20 tons of apples per acre at a total cost of $4,000 at 1990 prices (Hinman et al., 1992), or an average cost of $200 per ton orchard run. A typical orchard would yield 15 tons of fresh apples, three tons of juice apples and two tons of canner or peeler apples. However, the average return for the 1986-90 seasons was $309.44 per ton for fresh, $69.94 for juice, and $108.47 for canners or peelers. Thus, while the total return would exceed total cost, both juice and peeler apples would be sold at a substantial loss. However, without the admittedly low price returns for processed apples, the profitability of the individual grower would have been reduced by about 40 percent. If fresh returns covered all production costs, the returns from processed apples would be the only contribution to profit.

In contrast, the apple processing industry could not market its products as successfully in competition with the many other juice, pie and fruit snacks available if it were paying full cost of production for its raw materials as the fresh market must. Nor is this situation likely to change in most years. Since the demand for canners and peelers is static, their prices are unlikely to increase in real terms. The demand for apple juice is increasing, but in most years a ready supply of concentrate from many countries forces the U.S. juice price down to world market levels which are usually relatively low.

The structure of the market for processed apples is such that if processors can offer growers a price greater than the cost of harvesting and hauling, it will pay growers to harvest. Thus, in many eastern U.S. markets, processors publish weekly offer prices for canner and peeler apples of various sizes, for juice apples and for windfalls or drops. In Michigan, because of a unique mandatory state bargaining law, processors are required to set minimum prices for apples for processing in any season. Most apples are bought at or close to these minimums. However, given the fact that all processors have the same information on preseason inventory, demand conditions, expected crop size and world supplies, it is not surpris-

ing that price to growers for processing apples in Michigan corre-
lates highly with the national average in most years.

Increasingly, as fresh pack has been stretched through a longer
pack season, processors have been able to rely on culls from the
fresh packing line for raw materials throughout the season. These
culls may be bought on the open market for cash or may be con-
tracted from a specific warehouse to a specific processor. In some
cases, a grower may be a member of a cooperative packinghouse
and a cooperative processor, in which case that grower's cull fruit
will be automatically diverted to the processor, at the going price.
However, these automatic relationships have been known to break
down in periods of short supply.

In turn, processors generally must compete in a food market that
is usually supplied with many alternative fruit-based products. Pricing of the processed product must be tailored to the major competing products. For example, apple juice is generally the second most
popular fruit juice after orange juice. Thus, its price cannot move
too far out of line with that of orange juice without retail movement
being slowed. In addition, the product must be sold through a distribution system dominated by large retail supermarket chains, large
institutional feeders or large restaurant chains which have the power
to protect their marketing margins under most circumstances.

For example, between 1969 and 1984, the average wholesale
price of apple juice was approximately 3.8 times the grower price
for the equivalent amount of raw apples. The average wholesale
price of apple sauce was four times the grower price for the equivalent amount of raw apples. Retail prices for apple juice and applesauce were estimated at 5.6 times and 5.4 times farm value respectively in 1988-89 (Dunham, 1990). Data available for orange juice
suggest that the margin between retail and wholesale price averaged
about 22.2 percent of retail price. The retail margin for apple juice
was about 32 percent and for apple sauce, about 26 percent, both
higher than the margin for orange juice.

The future of the apple processing industry remains uncertain. It
seems likely that large, multiproduct food firms will become dominant in the marketing of processed apple products, as they have in a
wide array of manufactured foods. The small regional companies
whose main comparative advantage is in ready access to local raw

materials may have a difficult time surviving as independent sellers. Their options may include merger, integration further up the marketing channels or becoming multiproduct marketers themselves. In any case, the apple processing industry is likely to become increasingly food science driven, capital intensive and with need for fewer, but higher skilled employees in the processing operation. Just like the juice business, other segments of apple processing are likely to become more internationalized in ownership, procurement and marketing.

Chapter 9

Government Influence

Government has become a major influence in most business activities through its fiscal, monetary and trade policies. It can influence the supply and cost of credit, the level and incidence of taxation, and the strength and fairness of competition. While such policies are beyond the scope of this book, they can affect the success of any apple enterprise. In this chapter, we focus on more direct influences of government on agriculture, in particular on the fruit industry. We examine the rationale for these policies and some of their consequences. And, we look at ways in which government influence on the apple industry could be made more beneficial.

BY THE PEOPLE, FOR THE PEOPLE

In a democratic system, government's power arises when individual citizens cede some of their authority and their resources to a person or body to carry out specific functions for the common good. For example, a city might hire a town crier to warn of impending danger, to announce community activities, or to provide information of value to all citizens. Few people questioned the wisdom of having local government provide fire protection, police, roads, or basic education. Gradually, the same principle was applied to groups of cities and to even larger "communities" of interest. Thus, in the United States, governments exist with the power to tax and spend at the town, county, state, and national (federal) level. Limited government powers are also available to regional groups within states or across states, and to port, hospital and school districts.

Even in authoritarian systems of government, many actions are justified on the basis of the common good. There are few regimes so authoritarian that some provision is not made for citizen input to decisionmaking, either through an elite party membership or through direct petition to the ruler. While an authoritarian regime may be maintained by police power, even the most democratic system will require some sort of enforcement mechanism to ensure that taxes are paid, laws are obeyed, and public moneys are not misspent. Thus arises one of the many ironies of government, that those who voluntarily cede their authority to it, and, in general, welcome it, may also be forced to obey laws they do not like.

Aware of these dangers, democracies have attempted various formulas to restrain the powers of the governments they freely elect. One is definite terms of office, after which each elected official must face the judgement of the electorate anew. Another is the separation of powers between the executive, legislative, and judicial branches. The executive branch may propose activities but must go to the legislative branch for approval and funding. In turn, the legislative branch may pass laws but must rely on the executive branch to carry them out. The judicial branch, in turn, determines whether the actions of the other two branches conform to a written constitution or to precedent, and sits in judgement on disputes among citizens or between citizens and the government.

As governments at every level have become involved in more and more activities, citizens can often feel that they are being strangled by their own creations. As the layers of government become more numerous and their accompanying bureaucracies become more entangled, the average citizen can develop considerable paranoia about his or her own government.

For example, the newlyweds who decide to go into the orchard business find they have taken on a third lifetime partner, the government. They need a government permit to develop a piece of orchard land. They may rely on government for maintaining water, roads, or utility services to their property. If they wish to use imported rootstocks they will need a plant health clearance. Their choice of chemicals, rate of application, reentry into the orchard after spraying, etc., will be regulated. They must meet government rules on rates of pay, hours of work, health and accident insurance, and worker housing. The design and use of tractors, sprayers, and ladders and the storage of chemicals will be regulated. When they ship their fruit to a packinghouse, their rights to prompt payment and their security from unfair liens will be backed by government. Packing, grading, and inspection will be monitored by government. Their right to form a cooperative, to join a marketing order, or to be exempt from anti trust laws will be protected by government, which will also monitor the safety, sanitation and scheduling of the trucks bringing their product to market. Government will take the lead in testing the wholesomeness, nutrition, and freedom from harmful residues of final apple products. If our couple wish to market their products directly, they will have to meet a different set of food safety and sanitation rules. In the marketplace, they must meet various codes for advertising, labelling, pricing, or promotion. A sizable chunk of their membership dues to industry associations will go to monitoring, negotiating with, or attempting to influence various parts of government.

The young couple will be torn between regarding their government as a knight in shining armor when it protects their rights against large buyers, and as a dragon when it slaps them with fines for violations of an ever-lengthening code of business conduct. In the same breath, government will be seen as "our" government when it provides many useful services, and as "their" government when it imposes additional burdens on business. Citizens are caught in a dilemma of how to reduce government influence where it has a negative impact while maintaining or increasing its positive influence.

To a large extent, this dilemma has been created by the actions of the citizens who now deplore it. In a democratic system, any citizen or group of citizens has been free to define a problem as needing communal attention. If this does not lead to a consensus solution, the problem has been turned over to government to solve for the general good. Citizens have long known how to bring pressure on the executive branch to propose favorable solutions and on the legislative branch to fund and approve these solutions.

However, no single group of citizens can have anything but a partial and temporary victory. If the group is to tap government's powers, it must cede the same right to other groups. As additional groups learn to play the game, some of their requests may conflict directly with those already granted. In addition, the cumulative effect is a gradual and pervasive diminution of the rights and freedoms of the individual, and their transfer to an inflated government sector.

SATISFYING CITIZENS' WANTS

The satisfying of citizens' current wants becomes a pathway to power for both elected officials and civil servants. The politician can seek election or win reelection by looking after the special interests of constituents. In the legislature, coalitions are built not just around common goals but also over reciprocal goals: I'll support your special interest need in your district, if you support my special interest need in my district.

The Civil Service bureaucracy faces a more delicate game. The size of their budgets, number of employees and political clout depend on their ability to correctly forecast which interest groups and issues have longevity and which are transitory in nature; which are likely to gain legislative approval and which are not; which will create new spheres of influence, and which will cannibalize existing spheres of influence. For example, Departments of Agriculture have traditionally regulated on-farm use of agricultural chemicals. Vocal interest groups have called for a ban on all synthetic chemicals. If a Department of Agriculture opposes such policies and the politicians agree, that department may be allowed to retain its traditional regulatory role. However, if the politicians decide in favor of proponents of a ban on synthetic chemicals, the Department of Agriculture may have its regulatory authority taken away and given to a rival department.

Civil servants are also eager to mold pending legislation in a framework compatible with their existing structure. Units with many local branches may favor decentralized decisionmaking, while those with highly centralized programs will attempt to frame regulations accordingly. As the need for a major program fades, agencies become advocates for new programs to meet new needs, not the least being the maintenance of their sphere of influence.

In both democracies and authoritarian regimes around the world, the system of politically devised solutions to societal problems administered by a bloated bureaucracy is under intense strain and, in some cases, may already have collapsed. In centrally planned countries where the rule makers were distant in philosophy and access from the rule takers, faith in government has been severely shaken. Even in well-run democracies such as the Netherlands, many citizens are questioning how much longer one can add additional government involvement to the current, dysfunctional system.

There are many reasons why certain government programs may not work. The problem may have been incorrectly identified in the first place. Even if the problem was correctly identified, the compromises needed to get legislation passed may have led to a partial or inappropriate solution. For example, in food legislation, growers, packers, consumers, politicians, and bureaucrats may have quite different objectives. Often a bill is passed hurriedly and contains a mix of general sentiments, vague or ambiguous directions and some highly specific clauses to satisfy a particular interest group. The relevant government agency must then make detailed rules to operationalize the program. In the early years, it can be guided by legislative intent. But as politicians move on to other issues or other positions, the bureaucracy is frequently left as sole interpreter of intent and arbiter of detailed provisions.

Even perfectly good laws, honestly and fairly administered, can become ineffective over time as circumstances change. For example, when the Vietnam War highlighted the fact that 18-year-olds in the U.S. could be drafted into the armed forces and risk their life for their country, but could not vote, the voting age was lowered to eighteen. When a large proportion of auto accidents was traced to teenage drinking, under federal pressure the legal minimum drinking age was raised from 18 to 21. Now 18-year-olds could vote on drinking laws but not legally drink. Of course, drinkers under 21 continued to drink, but illegally. This latter law shows the difficulty of legislating personal behavior. However, other laws can be rendered just as ineffective if they are difficult to monitor, weakly enforced, are not supported by the public, are costly to comply with, or appear counterproductive.

Perhaps the biggest single problem with the democratically legislated approach to government is that the cumulative effects may be the opposite of what individual laws intended. The goal may be to make life safer, happier and healthier, to reduce the risks in operating a business, or to alleviate the burden of social or medical problems. However, the actuality is often very different. While emission controls attempt to reduce the pollution from each automobile, highway building programs stimulate the number of automobiles on the road. While numerous welfare programs treat the effects of family breakdown, easier divorce laws facilitate family breakup. While civil rights laws encourage equity in education, housing and employment, the relative status of many minorities deteriorates.

GOVERNMENT IN AGRICULTURE

Agriculture has been deeply affected by the long buildup in government involvement and by the eventual unravelling and self-defeating nature of the measures taken. Government has long had a special interest in helping agriculture.

The era of large-scale and continued involvement of government in agriculture began during World War I when governments in many countries adopted special controls and incentives for agriculture to preserve food supplies during the hostilities. Government has found a number of different rationales for paying special attention to agriculture. Food was a basic need, and food supplies were of strategic importance in a turbulent world. Farmers were a major share of the population in the 1920s and 1930s, and despite their diminished numbers, remain an important voting bloc in many districts in the 1990s. Farmers were considered to be of social importance because of their close communion with nature and their key role in many rural communities. The farm business was particularly deserving of emergency government relief because of its vulnerability to "Acts of God," such as excessive heat, cold, floods, drought, pests, or insects. More recently, farmers' importance to the environment has been recognized because of their role as stewards of the nation's land, air and water.

In the more-developed countries of North America, Western Europe, Asia, and Australasia, as general prosperity increased and farmer numbers shrank, governments were willing to devote increasing dollar amounts to farm programs. Farmer incomes, on average, lagged behind the incomes of nonfarmers. This gap, which could be interpreted as a market signal to marginal farmers to seek alternative employment, instead was interpreted by many governments as a signal to attempt to maintain farmer numbers by artificially stimulating incomes. For reasons described in detail elsewhere (e.g., Pasour, 1990), decades of farm programs have failed to help marginal farmers.

Another popular rationale for supporting farmer incomes was to ensure self-sufficiency in food and other agricultural production. On its face, the logic is attractive. If British farmers could grow all their apple needs, why not stimulate British apple production and end imports from France. However, on further reflection, it can also be shown that such a policy would be bad for Britain and possibly even bad for British farmers. As shown in an earlier chapter, growing a ton of apples requires considerable resources in land, water, energy, labor, capital, and management. Applied to a favorable natural environment, such resources could produce apples as effectively in Britain as in France. However, applied to an environment currently deemed more suitable for cereals or dairying, Britain would both lose the potential output of cereals or dairy products and have to use more resources in growing the apples than if they were imported from France.

Pushed to its extreme, a self-sufficiency policy would lead a northern country like Finland to grow its own bananas or citrus fruits. To do this, it would have to utilize a tremendous amount of resources to create an artificial environment in the harsh Finnish climate. In such extreme cases, governments can usually see the folly of total self-sufficiency, but countries have greater difficulty in resisting self-sufficiency programs where the marginal differences are not as clear-cut.

After 60 to 70 years of government initiatives at local, state, and national level to help farmers, the evidence is convincing that the help has been counterproductive. In the case of apples, the greater the help, the worse shape the apple industry is now in. The more protection a national apple industry has received, the less competitive it may be in the world market.

CANADIAN APPLE INDUSTRY ASSISTANCE PROGRAMS

Government intervention in the apple industry throughout the world is so pervasive and varied that even a summary would require a book in itself. A detailed study by Lusztig (1990) on government programs in the British Columbia, Canada apple industry is a representative example of the types of programs attempted in other countries. Apple production in British Columbia is limited primarily to the Okanogan Valley, a long way from most major markets in eastern Canada or the United States. Lusztig was writing in his role of Commissioner to the Commission of Inquiry into the British Columbia Tree Fruit Industry.

Lusztig found that in the sixteen years, 1974 to 1989, about 3000 growers in the B.C. tree fruit industry received over $357 million (1988 Canadian dollar equivalents) in major provincial and federal government expenditures. This amounted to an average of $115,000 Canadian per grower, and about $12,000 per acre. "From 1974 to 1980, direct government payments represented 21 percent of apple returns growers received from the market. From 1981 to 1988 they accounted for 37 percent of market returns. With indirect payments included, the amounts are 25 and 46 percent respectively."

Lusztig described some of the programs included in those calculations, including subsidized loans for land development and the adoption of new technology and environmentally sound practices, loans at half the bank prime rate for orchard renovation, and property tax exemptions and reductions. Over half of payments in the 1974-89 period were under the Farm Income Insurance program established by British Columbia in 1973. Under this program, a representative cost of production each year was negotiated between the B.C. Fruit Growers' Association and the Ministry of Agriculture and Fisheries. If total returns from the market and from other government payments fell below the estimated cost of production, the B.C. government paid growers a subsidy of half the difference. Lusztig noted that estimated costs had exceeded returns in every year of the program's operation, casting doubt on the process by which costs are estimated. He also criticized this method of subsidy as a disincentive to improve quality, increase market returns, or reduce costs. In addition, the benefits of the program were likely to be capitalized into land values, raise land prices to new entrants and feed back into the program as higher costs, thus fuelling another increase in subsidy payments.

Other B.C. support programs included an agricultural land reserve which has been used to prevent orchard land being bid away for nonagricultural uses, a provincial subsidized crop insurance program, and an Agri-Food Regional Development Subsidiary Agreement which funded projects that would enhance productivity, maintain and improve soil, land and water resources, and provide assistance in processing and marketing. A joint federal provincial agreement called the National Tripartite Price Stabilization program provided payments to particular commodities, including apples, when price fell below a specified level. Growers contributed a premium to this fund which was matched by federal and provincial contributions, and paid none of the administrative costs. Two other important indirect supports mentioned were the Tree Fruit Extension Services and the Summerland Research Station.

Lusztig commented that

> With few exceptions, programming, whether Federal, Provincial or joint, generally appeared to be a collection of somewhat ad hoc responses to perceived shorter-term financial needs. Convincing evidence of a systematic approach to support measures and programmes is lacking and there is no indication that a strategic plan designed to help the industry become more competitive, innovative and self-sufficient underlies governmental backing. Not surprisingly, therefore, despite fairly sizeable annual commitments by government and an impressive array of program initiatives (including private sector instruments of delivery), the problems facing many growers remain acute and have changed but marginally over the years. (p. 63)

Lusztig concluded on behalf of the Commission that the way ahead for the B.C. tree fruit industry was to gradually wean the industry away from government subsidies and allow the chill winds of the free market to blow through and reinvigorate it. However, as his study was being completed, Canadian growers were in no mood to accept such medicine. They had become enamored of a suggestion to establish a National Apple Agency with supply management powers over producers of fresh apples, and the ability to restrict or prohibit imports. Despite criticisms that such a quota scheme would lead to further inefficiencies in the Canadian apple industry, impose additional costs on consumers and society and cause international trade frictions, including retaliation against B.C. exports, the supply management scheme continued to enjoy heavy grower support throughout Canada. This is a classic example of an interest group attempting to co-opt government powers to gain a perceived advantage for the group.

The apple industry in many other countries has been successful in winning government favors for its members. As the leading deciduous fruit in almost all the more-developed countries, the apple industry has been heavily subsidized in many cases. As Lusztig's description of the major Canadian government programs showed, government interventions have taken many forms. The financial and economic effects have not been readily transparent.

COMPARATIVE APPLE SUBSIDIES

To deal with this issue, economists have developed a methodology for transforming various schemes into producer subsidy equivalents (PSEs). Essentially, the effects of programs of various kinds can be converted into the equivalent dollar value of the benefit to growers. The dollar value of property tax relief, subsidized credit, below cost insurance and other programs can be summed to permit an objective comparison of the level of subsidies between commodities and across industries and between countries.

As a contribution to the discussions on reducing agricultural subsidies in the Uruguay Round of the GATT negotiations, the Organization for Economic Cooperation and Development measured PSEs for selected apple producing countries (OECD, 1991). The PSEs included:

1. All measures which simultaneously affect producer and consumer prices (e.g., market price supports).
2. All measures which transfer money directly to producers without raising prices to consumers (e.g., direct payments).
3. All measures which lower input costs (e.g., subsidies to capital or other inputs).
4. Measures which, in the long term, reduce costs but are not received directly by producers (e.g., general services).
5. Other indirect support (e.g., tax concessions).

Producer subsidy equivalents were reported in three forms:

A. Total value of transfers to apple production.
B. Value of transfers per metric ton of apple production.
C. Transfers as a percentage of total value of production (including transfers).

Percentage PSEs are summarized for the average of the years 1979-80 and 1988-89 (Table 9.1). For these major apple producing countries which normally account for more than one-third of world apple production, one dollar in every five received by producers had come from government subsidies. Subsidy rates for all crops were even higher than those for apples, and increased substantially during the 1980s. Subsidy rates for apples rose in every country but New Zealand. Countries which had relatively high PSEs in 1979 and 1980 also had relatively high PSEs in 1988 and 1989. Of the countries which provided higher PSEs to apple growers than to all crops in the earlier period, only Australia and New Zealand continued to do so in the later years. The actual dollar PSE per metric ton of apples in 1989 ranged from $10 in New Zealand to $536 in Japan. Next highest PSEs were $297 per ton in Norway and $117 per ton in Australia. The OECD data probably understate the level of subsidy to apple growers because minor programs and local or regional subsidies are occasionally excluded.

The OECD data included the subsidies due to direct and indirect payments and those resulting from various forms of trade protection. The types of subsidies used varied widely by country. Australia and New Zealand relied primarily on market price supports through a two-price system of charging higher prices in the domestic market and lower prices for exports. Japan's primary market supports resulted from total exclusion of cheaper imports. Norway and Sweden kept domestic prices artificially high by various curbs on imports. The European Community relied about equally on market price supports and reduction of input costs. Its market price supports resulted from tariffs which varied seasonally, being most prohibitive in the fall. Most of the reduction in input costs derived from member country measures rather than European Community programs. In Canada, over 40 percent of subsidies came from provincial programs, and close to 30 percent from deficiency pay-

TABLE 9.1. Percentage Producer Subsidy Equivalents, Apples and All Major Crops, Selected Countries, 1979-80 and 1988-89

Country	1979-80		1988-89	
	Apples (%)	Crops (%)	Apples (%)	Crops (%)
Canada	19	14	25[1]	28
United States	7	9	12	29
Australia	17	4	31	11
New Zealand	13	3	8	7
Japan	32	75	36	88
EC[2]	18	38	29	44
Norway	53	71	49	81
Sweden	28	26	26	44
Average	18	28	26	46

SOURCE: OECD (Organization for Economic Co-Operation and Development). 1991. *The Apple Market in OECD Countries*. Paris, France.

[1]Data for 1987. [2]EC-10 until 1985, EC-12 from 1986.

ments. The major sources of subsidies in the United States were reduced input costs (primarily fuel tax exemptions) and general services, including research, extension and teaching, and soil and water conservation.

JAPANESE APPLE INDUSTRY ASSISTANCE PROGRAMS

While measurements of PSEs is a useful approach for comparative or summary purposes, it does not indicate the many schemes which governments continue to employ. For example, while the PSE for apples in Japan was less than half of the PSE for all crops, that did not mean that government intervention was minor. Protectionist policies for rice led to huge excess supplies and government incentives for farmers to transfer land to intensive crops such as fruits or vegetables. These crops, in turn, were stubbornly protected from external competition by high tariffs, quotas, and, in the case of apples, outright import bans. Returns to apple growers were boosted both by protection on apples and by protection on other substitute fruits. Under external pressure, these barriers were gradually being removed in the late 1980s and early 1990s.

The fruit industry in Japan was the focus of selective expansion and production diversion programs of the Japanese Ministry of Agriculture, Forestry and Fisheries, including special government assistance, financing for cooperative processing factories, and extension services to farmers (Australian BAE, 1988). The government attempted to encourage an increase in tree numbers by providing subsidies for production machinery, for sorting and packing equipment, for the planting of dwarf varieties and for one-third of the cost of installing new storage facilities. While these supports assisted the industry in replacing less desirable Delicious and Golden Delicious strains with newer varieties such as Fuji, Tsugaru, and Mutsu, total acreage planted and tonnage produced changed little. Per capita demand showed little response to rapidly increasing consumer incomes (Heydon and O'Rourke, 1981). Thus, total production in 1990 was about the same as in 1970. The Ministry of Agriculture, Forestry and Fisheries, through prefectural governments, controlled plantings within a static national target.

EUROPEAN COMMUNITY APPLE INDUSTRY ASSISTANCE PROGRAMS

The world's leading apple producer, the European Community, supported apples using selected features of its common agricultural policy, the notorious CAP. As a customs union, the European Community attempted to maintain free trade between member countries while discriminating against nonmembers through a common external tariff which, in the case of apples, varied seasonally. However, the concept of a common internal market was breached in 1969 when the French unilaterally devalued their currency. It became unsustainable when the currencies of many countries were allowed to float freely in the 1970s.

The principle underlying the CAP was to annually set community-wide minimum or target prices for each commodity to be supported. Initially, these were set in terms of a "unit of account," equivalent to one U.S. dollar. Since all member currencies had a fixed exchange rate with the dollar, the equivalent support price in French francs or German marks was fixed for the year (Gardiner, 1990). Under floating exchange rates, support prices in national currencies could vary daily, and did. For example, suppose the price for 10 kilos of apples was initially set at four French francs or two German marks equal to one unit of account. If the value of the French franc dropped to where five French francs equalled two German marks, French growers would now get less support payments in real terms. For each four francs worth of apples they shipped to Germany, German importers would only have to give up 1.6 marks. German growers would find French apples now 20 percent cheaper and would feel that they were being undercut in their own market. To prevent these market effects, the European Community established monetary compensatory amounts (MCAs). There were border taxes on exports, or subsidies on imports of weaker currency countries, and subsidies on exports or taxes on imports of stronger currency countries. These MCAs were based on the difference between the actual exchange rate and the rate designed to maintain agricultural price parity, thereby dubbed the "green" rate. Since the administrative adjustments could not keep pace with the volatile nature of the real exchange rates, multiple green rates proliferated as countries attempted to cope with short-term advantages or penalties imposed on their growers. Much tinkering with the MCA system failed to eliminate the trade distortions. It became clear that a stable community-wide support price would have to await a common Community currency, not now expected until the twenty-first century.

While the execution of a common internal support price was thwarted, its goal of maintaining European Community prices above world market prices was not. Each year a "reference" price was established for each product. If the price of competing imports delivered to major internal markets fell below that reference price, a countervailing duty would be added to subsequent shipments (Ritson and Swinbank, 1984). These duties were used to fund part of the cost of the common agricultural policy.

The common agricultural policy for fruits and vegetables was maintained through compulsory standardization of products, producers' groups, price and intervention arrangements and rules for trade with nonmembers (European Community Commission, 1982). Standardization of grades facilitated volume control by allowing elimination of lower quality products. Producers' groups were subsidized to provide discipline in normal production and marketing and as agents for Community policies. For apples and other major fruits and vegetables, the Community established "basic" prices for "pilot" products, usually normal producer prices for class I products. It set "buying-in" prices at about half of basic prices. If market prices fell to buying-in price levels, or no market could be found, producers' groups could pay members a "withdrawal" price, usually a high proportion of the buying-in price. Rules for trade included the countervailing duties on imports described above, and refunds on products exported.

The use of withdrawal for perishables such as apples aroused widespread criticism. The original intent of the CAP intervention system was to prevent sharp declines in market prices by buying in temporary excess supplies of storable commodities which could later be resold at higher prices. In theory, such an intervention system made sense for a commodity like butter which could be frozen in relatively good condition for months. Of course, the very buildup in stocks overhanging the market tended to prolong the period of depressed prices, the need for continued intervention, and the swelling of the notorious butter mountains. In the case of a product such as apples, where the passage of time causes predictable and inevitable product deterioration, resale out of intervention was rarely profitable.

Apples from intervention were disposed of (1) to charitable organizations and other "social" institutions, (2) to the distilling industry, and (3) for animal feed (European Community Commission, 1984). In some years, such as 1981-82, half of all withdrawn apples went unused. Use of withdrawal apples for juice was suspended in 1984 because of cost. Also, rather than encourage further use of the withdrawal system, the European Community occasionally provided subsidies for the uprooting of orchards or for the replacement of obsolete varieties.

The United States consistently had the lowest producer subsidy equivalents for apples among major producing countries. Most of this consisted of the estimated share of benefits received by apple growers from general schemes for all farmers such as fuel tax exemptions, research and extension, and government-subsidized irrigation networks. In most years, apple growers preferred to take their chances with the market rather than endure the constraints of the type of government programs in operation for grain and oilseeds. However, in years of economic distress, growers occasionally sought limited assistance through government purchases of processed products, increased use of fresh apples in school lunch programs, or purchase of surplus stocks during the Alar shock of 1989. In general, these ad hoc interventions came too late in the marketing year to be of much benefit to most growers.

NONGROWER ASSISTANCE

Not all government interventions occur at the grower level. As previously mentioned, the Japanese government has assisted with modernization of packing and storage facilities, the European Community has assisted cooperatives and producer associations, and the U.S. has purchased processed apple products for its school lunch and food distribution (welfare) programs. However, government intervention at the farm level inevitably affects the supply, quality or price of raw materials going to packers or processors. Accordingly, packers and processors have been active participants in influencing government intervention, both at the farm level, and also, where necessary, at the processor level. For example, under the EC's common agricultural policy, about 70 percent of price support expenditure was channeled through the food industry for eventual distribution to farmers (European Community Commission, 1983). Processing industries received both direct assistance, such as investment aid, and indirect help such as waivers of customs duties and other taxes. Processors have also been the beneficiaries of intervention buying, consumer subsidies and export subsidies.

Processors have also been viewed as a vehicle for economic development in rural areas. Governments all over the world have followed similar logic in believing that the existence of adequate processing facilities in an area can be the catalyst for expansion of farm production, an increase in farm incomes and a boost to rural development. However, in a review of such policies published by OECD, Roucaud and Yon (1983) argued that

> An establishment that does not reflect the independent choice of a good businessman is an artificial one. Because of various difficulties in the development of the regions there is sometimes a temptation for the authorities to try to create an agrofood industry on their own initiative, without calling on entrepreneurial partners; for example, through an agricultural cooperative, which can provide a policy excuse for the use of public money for rural development. If, however, the spirit of enterprise and managerial capacity is lacking, a firm that has been created artificially will inevitably run into difficulties and this always leads in the end to the elimination of ill-founded economic activity.

On the other hand, one can point to situations where government intervention made a difference to the existence or progress of an agricultural industry when other business conditions were favorable. Government authority has permitted the damming of major rivers, despite losses of whole communities upriver of the dams and loss of wild stream flows below the dams. Government funding has permitted the long-term investment in irrigation delivery systems from these dams and the delivery of below cost water to farmers. Since these farmers already had good soil, climate and other favorable conditions, they have been able to grow apples and other fruits competitively for world markets. Many of the best apple growing districts in the world owe their existence to government-subsidized irrigation.

The crux of the matter seems to be that government can rarely discipline itself in its interventions, even in cases where it recognizes that policies have failed. For example, the European Community on many occasions has tried to reform its common agricultural policy on the rationale that its existence was vital to the survival of the Community. At the same time, its own executive branch, the European Commission, in reviewing the CAP, talked in a 1989 report of "a flagrant paradox: mounting agricultural expenditure and plummeting farm incomes" (p. 53). The Commission pointed out that in the years between 1975 and 1988, CAP price supports in real terms rose by more than 160 percent, while gross domestic product in the Community increased by only 32 percent and the volume of agricultural production by over 25 percent. "Two-thirds of the Community budget (and in some years significantly more) were steadily poured into agricultural market support, to the detriment of other Community policies, which had to be pruned to a minimum because of the tight budget situation" (p. 53).

GOVERNMENT NONPRICE PROGRAMS

If this paradox of government intervention persists in an area such as prices or incomes that can be easily measured, it is reasonable to ask what the chances are for rational government policies in some of the murkier areas in which government has now become active. For the apple grower, these areas include information, all aspects of chemical use, worker relations, food safety, freedom to organize collectively, marketing methods, and the cumulative effect of government programs.

Government has become the major dispenser of information that is used by the apple industry, both in day-to-day decisions and in long-term planning. Governments conduct periodic farm censuses, develop crop forecasts, produce situation and outlook analyses, monitor prices and shipments, and track marketing margins, freight rates and other indices of industry health. The benefit of this information is usually not questioned. However, on closer examination, it is not obvious that government-funded information always works to the benefit of growers. For example, in the United States, apple prices are reported daily in great detail at shipping point level, but only monthly for one variety at retail. Thus, buyers can get daily information on shipping point supply and price, but sellers get virtually no public information on retail price and movement. Crop forecasts also occasionally engender controversy, particularly if a large crop is being forecast. Growers are sometimes concerned that larger buyers are better able to utilize such information in advance planning than are the many small sellers. As cost pressures force government to reassess the information services they provide, growers and shippers may be forced to evaluate the benefits of each of the services provided.

Government has also become deeply involved in the supervision of chemical use in agriculture. From the publication of Rachel Carson's *Silent Spring* in the 1960s, concerns about the adverse effects of chemicals have triggered a series of ever-widening legislation and expansion of agencies to administer that legislation. In the United States, the Food and Drug Administration, the Environmental Protection Agency, the Occupational Safety and Health Agency, and federal and state departments of agriculture play overlapping roles.

Much of the policy on agricultural chemicals has derived from fear and emotion rather than the careful data collection and analyses that such all-encompassing and expensive programs require. For example, the 1958 Delaney amendment to the Federal Food, Drug and Cosmetic Act of 1938 permitted zero tolerance for chemicals that could be shown to be capable of producing cancerous tumors in test animals (Paarlberg, 1980). Such a tolerance dominated research and decision making on chemicals for a quarter of a century. Gradually, the reality emerged that chemicals are not a foreign substance, but the basic building blocks of animals, plants and minerals. Chemicals synthesized by man continued to suffer a universal stigma, but work by Bruce Ames and others (e.g., Ames and Gold, 1990), has demonstrated that plants and animals generate natural chemicals for many purposes that can be as hazardous to humans as synthetic chemicals.

The general government approach to control of chemical use has been to classify some chemicals as good or bad, and to ban the bad; and to specify appropriate concentrations for specific uses, like a pharmacist meting out cough medicine to a child. Unfortunately, such centralized decision making has usually been time consuming and has rarely fitted the many and varied circumstances in which a chemical is to be used.

The three major aspects of chemical use of concern have been the potential hazard to workers, the potential effect on the environment (land, water, air) and the danger of residues to consumers. The initial focus of government programs was to make rulings based on expert analysis and attempt to enforce those rulings by punitive measures. Gradually, this approach in agriculture has given way to a more cooperative, educational approach. It is not a particular chemical which puts workers, the environment or consumers at risk, but how carefully and appropriately that chemical is used. Only the individual apple grower or warehouse operator is in a position to weigh the benefits and costs of using a particular chemical, of fine-tuning the dosage to the problem being treated, and of adequately supervising chemicals, sprayers, equipment, storage and disposal.

Government must learn to trust apple growers and other farmers to be motivated to meet worker safety, environmental and food safety concerns, and apple growers themselves must show themselves worthy of that trust. It must provide incentives to the responsible growers and disincentives to the irresponsible. Otherwise, apple growers must face even more stringent regulation in the future. Advances in understanding of risk assessment and risk management offer the hope that regulation in the future can show more common sense.

Worker relations will continue to be of major interest to growers. For a long time, agriculture argued successfully that its special nature justified its exemption from many labor laws, but that position has steadily been eroded. Because of its intensive use of labor, the produce industry has been particularly resistant to measures that might increase the cost of labor. However, over time, governments have included agriculture in many of the regulations which applied to other industries and have added other measures directed specifically at agricultural operations. Apple growers must make payments for social security, unemployment, and workers compensation (in the case of injury). They must shoulder much of the responsibility for training and supervision of workers in safe practices for operating machinery, equipment and chemicals. They must police reentry rules after spraying, monitor storage of chemicals, and keep records of chemical use. They must provide adequate housing for migrant workers. The standards for this housing tend to rise as general standards of housing in the nonfarm population rise.

In most cases, these regulations impose additional fixed costs on an orchard which must primarily be offset by higher yields. However, some orchardists have turned regulations to their advantage by using them as an opportunity to make workers better trained and more productive. For example, faced with responsibility for chemical misuse by employees, many U.S. orchardists have learned Spanish to better communicate with their Hispanic workers.

PROGRAMS TO CONTROL CHEMICAL USE

Government intervention has tended to be most intrusive in the control of chemical use by orchardists. Fruits offer a concentrated and attractive food source for insects, mites, viruses, and bacteria of many kinds, both during the growing season, in storage and in the marketing system. Left unchecked, populations of insects, etc., grow very rapidly and can devastate an entire orchard in a matter of days. For example, if left untreated, a codling moth infestation can sting every apple in a block, rendering most unsalable in the fresh market.

The more intrusive the production system, the more extensive the battle plan an orchardist must have in order to stay ahead of the many natural enemies. This battle has been made more difficult by the ability of insects, etc., to build up natural resistance. That fact led scientists in the 1960s and 1970s to promote the system of integrated pest management. The main elements of IPM were (1) scouting the growth of pest populations, (2) withholding spraying until pest damage had reached an economic threshold, (3) spraying at the most strategic points to eliminate the maximum number of pests at one time or to do most damage to the next breeding cycle, (4) encouraging populations of beneficial insects, and (5) using other nonchemical techniques to reduce the number of sprays used and to delay the build-up of resistance to pesticides. IPM was successful in reducing both the quantity and cost of chemical use on apples in certain situations but not in eliminating chemical use (King and O'Rourke, 1977).

In the 1970s and 1980s, however, more and more chemicals faced limitations or outright elimination by the EPA or FDA based on perceived hazards to worker health, the environment or food safety. Apple growers found themselves frequently in an adversarial position relative to these government agencies. The regulatory climate did not only threaten to take away existing chemicals that were pivotal to pest control, but it slowed the approval of newer, and supposedly safer, replacements. In the eyes of third parties, government presented itself as the defender of consumer health, worker safety and a clean environment, and farmers were painted as recalcitrant and callous polluters. Farmers usually lost the battles with government over retention of specific chemicals.

In the late 1980s, the movement towards a "sustainable" agriculture gained adherents, although there was much disagreement over what was meant by sustainable. To some, sustainability meant maintaining the productivity of the earth by halting soil erosion, preventing air or water pollution, and adapting current techniques as new knowledge became available. To others, sustainability was only possible if mankind retreated to the lifestyle of prehistoric ancestors, not disturbing the soil with tillage and absolutely not using chemicals in the battle with pests. To the latter, the modern intensive fruit farmer was an enemy of nature, and any means were justified to disrupt modern farm practices.

In a comprehensive review of the sustainability debate, Tweeten (1991 p. 9) argued that "An environmentally sound agriculture is a worthy objective widely supported by Americans, but the economics and policies for (it) remain in a formative stage." It is particularly unrealistic at this time for perennial crops. The available studies for field crops suggest that costs would rise and yields would fall. If use of all chemicals (including pesticides, herbicides and fertilizers) was ended, aggregate production would fall so sharply that prices would rise. Growers' total revenue would increase, but so would consumer prices, very substantially.

An intermediate stage being promoted for fresh produce such as apples—organic farming—faces many difficulties. Many consumers do not understand that "organic" normally means avoidance of synthetic chemicals, but is in no way guaranteed to lead to lower total chemical use, less total toxicity or lower residual chemicals. In fact, in the case of apples, the total volume and cost of chemical control usually increases. In addition, the demand for "organic" produce appears very sensitive to the price gap between conventional and organic produce. As the supply of organic produce increases, the premium for growers disappears, and the incentive to undertake the greater management requirements of organic farming is reduced.

In this philosophical struggle, government finds itself in the middle, assailed on one side for putting the environment before farmers, on the other side for favoring farmers at the expense of the environment. It is difficult to predict what the eventual outcome of this conflict may be. Almost certainly, government will not easily give up its role of monitor and judge of chemical use in agriculture.

The food safety debate involves a similar array of viewpoints, ranging from those who simply ask for a little more caution in current practices to those who seek zero hazards in the food they eat. It is complicated by the scientific uncertainty surrounding the risks involved in food and its constituents. Pressures are placed on government from all sides in the debate in order to win rulings favorable to one side or another. Under pressure of elections or other deadlines, politicians often make decisions before the full implications of those decisions are known.

Governments have frequently injected themselves into the marketing system for apples and other products. In many centrally-planned and less-developed countries, a Ministry of Supply or its agencies have monopolized purchases. In cases of limited foreign exchange, apples have often been considered a luxury good and have been excluded from purchases with serious disruptions to world markets. Many countries have required apples to be sold through a government marketing agency or only through traders approved by the government. In the U.S., marketing orders have been employed in many fruits and vegetables, but only rarely, on a state basis, in apples. Limited marketing orders have permitted growers to tax themselves to fund promotional programs, research and market development. In other countries, apples have faced a sales or export tax, sometimes mitigated by rebates to selected ports, markets, etc.

Governments have also interfered in markets more directly by fixing prices or marketing margins to the consumer. In the U.S. in the 1970s, price controls were imposed on processed products, but not on fresh. As a result, retailers reacted by expanding their sales of fresh products and lessening their use of processed products. One unintended consequence of this government action was to contribute to the reversal in the popularity of fresh and processed fruits which occurred in the mid-1970s. In other countries, government price-fixing on produce has led to a black market in the higher quality products and wide availability at the official price only of product of mediocre quality.

Any catalog of government initiatives can be matched with an equally long list of failed initiatives and unintended side-effects. The previous assessment of price and incomes policies indicated widespread failure and unintended results. No estimate has been made of the aggregate effect of the many government programs in operation in many different aspects of the apple industry. Comparable analyses that have been done, for example, by Pasour (1990) on agricultural policies in general, come to rather bleak conclusions. Pasour argues that government farm programs that operate through regulation or through subsidies are generally doomed to failure. He uses words such as ineffective, inconsistent, anachronistic, and anti-competitive. In addition, he argues that "there is a systematic bias toward expansion of the role of government in agriculture as in other areas" (p. 58).

The move to privatization of many government functions which has swept through many centrally planned economies has not been as eagerly accepted in more democratic societies. There is still faith that democratic expression which can be translated into government programs will actually deliver what was promised, even though the evidence to the contrary is very strong.

A December 24, 1991, story in *The New York Times* reported on continued growth in federal regulation in the United States despite the efforts of the White House Council on Competitiveness headed by Vice President Dan Quayle. In the three years following the end of the Reagan presidency, the number of proposed federal regulations had risen by one quarter. President Bush had signed many pieces of legislation that would lead to new regulations, including the Clean Air Act, the Americans with Disabilities Act and laws on airplane noise, oil spills and food labelling. The *Times* reported that "the annual estimated cost of environmental regulations is now $123 billion and is expected to reach $171 billion by the year 2000. The Clean Air Act alone will cost industries at least $25 billion a year" (p. A1).

Because of the many ways in which the apple industry's activities are intertwined with government, the future progress of the debate on the role of government will continue to be the source of much irritation and considerable interest.

Chapter 10

International Trade

International trade has been an integral part of agriculture for millennia. From the beginning of recorded history, traders have been willing to cross mountains, rivers, deserts and oceans and to brave the hazards of weather, wild animals and hostile natives in order to satisfy people's yearning for goods from other lands. The great empires of Greece, Rome, and China drew vast quantities of grain and other produce from their more fertile colonies. The European empires which developed after Columbus' voyage to the new world in 1492 relied heavily on their colonies for supplies of agricultural and other raw materials.

For 200 years until the Civil War period, the North American continent was a major supplier to Europe of nonperishables such as grains, cotton and tobacco. It was only in the twentieth century that the journey from North or South America, South Africa, or Australia became short enough by steamship to permit long-distance shipment of more perishable items such as meats, fruits or vegetables. Even then, the volume was not large. However, that volume has continued to grow throughout the twentieth century.

THE GROWTH OF INTERNATIONAL TRADE

International trade has suffered periodic setbacks, most notably during the disruptions caused by the hostilities of World War I (1914-18) and World War II (1939-45) and by the subsequent damage to ports, vessels, warehouses, equipment, and trading infrastructure. Between the wars, many industries, including agriculture, laid the blame for recessionary conditions on imports. This spawned a rash of protectionist measures among major trading countries that put a temporary damper on trade expansion. Despite the fact that protectionism, like sin, retains a strong attraction, the major trading nations have fought hard to resist it in its many forms in the second half of the twentieth century.

Despite the many obstacles it faces, international trade persists because of a few simple principles. One is the "law of comparative advantage." In its most basic form, this law states that if two countries produce the same two products, even though one country is more efficient in the production of both products, it will still be worthwhile for the countries to trade if the first country has a comparative advantage in the production of one product. An example for the fruit industry is the situation where Canada and the United States both produce apples and sweet cherries. Even though the United States has an absolute advantage in the production of both apples and sweet cherries, since it has a comparative advantage in the production of sweet cherries, both countries can be better off if Canada imports much of its sweet cherry needs from the United States and exports apples to the United States.

Trade in agricultural products follows naturally from the law of comparative advantage. The production of most agricultural products is optimal in a limited range of soil and climate. Regions and countries rapidly recognize in which products they can most profitably specialize and which products it may be best to purchase or import. Clearly, one can understand why a country in a temperate region would grow apples and import its bananas from tropical regions. It is less obvious why two apple-producing countries would trade apples. However, producing areas can specialize in different varieties, storage regimes, timing of sales, packaging or merchandising devices to differentiate their products. The law of comparative advantage applies equally to these subcategories as to broader product categories.

The actual volume of trade between two countries is determined by supply and demand conditions in each country. In its simplest form, trade will take place if the exporting country can earn more after paying transportation costs by selling a lot in the export market than by selling that lot in the domestic market. As more lots are shipped from the exporting country to the importing country, the price in the exporting country rises and the price in the importing country falls. At some point, as the price in the importing country falls and in the exporting country rises, the gap becomes less than transportation costs, and trade will stop. Thus, the forces of supply and demand in both countries provide an automatic adjuster to the volume of trade that will take place.

However, that automatic adjustment process is rarely reassuring to producers in the importing country. They dislike imports because of their price depressing effect. They are easily persuaded that imports are benefitting from some form of unfair subsidy, being fully aware that any limitation on imports will strengthen domestic price. Consumers in the exporting country also will face higher prices as exports expand. However, it is only in extreme cases that this effect will become so widely known that consumers will call for a limit on exports.

Mackey points out that the major economic factor affecting the growth of trade over time is growth is per capita income in the importing country (USDA, 1965a). In the standard sense, for normal goods, any increase in income will lead to an increase in demand. Income increases also stimulate imports in another way. As the income of the individual consumer rises, he or she will tend to buy a wider selection of products, including novelty or specialty items. Import items are most likely to fit this niche.

At the firm level, export success is dependent on the same factors that govern success in the domestic market: being competitive in price, quality, service, and value. As in the domestic market, niches exist for many combinations of value ranging from premium-priced, high-quality products to cut-price items for the low-quality market.

The major difference involved in international trade is that buyer and seller operate under different governments and different legal systems. Governments have been involved in monitoring international trade for many centuries and are increasingly attracted to introduce additional regulation. Citizens will accept restrictions on foreigners that they would not tolerate on themselves. Much of the movement towards freer trade involves efforts to identify these restrictions and gradually disarm them. Essentially, traders must find ways of living with and bearing the costs of trade regulations and must be able to cover these costs if trade is to take place. The growth in world trade despite all these obstacles is a testament to the many benefits that trade brings. Consumers have a powerful desire to buy many and different products from many lands. Exporters and their suppliers, importers, and the distribution system, all benefit from handling the traded products. Countries in aggregate, by expanding their exports, can earn the foreign exchange to enable them to buy a wide array of imports in return.

Trade in apple products has grown steadily since World War II, primarily for fresh apples (Table 10.1). Growth of apple juice trade was discussed in Chapter 8. Because apples are produced in many different countries and because per capita consumption tends to be highest in traditional producing countries, only about one-tenth of production currently enters international trade. However, as shown in Chapter 1, the proportion of fresh apples traded is by no means uniform. While countries like Germany, the United Kingdom, Norway, Sweden, and Taiwan are major importers of fresh apples, France, Italy, Hungary, Argentina, Chile, New Zealand, and South Africa are major net exporters.

Codron (1990) analyzed the doubling of exports of deciduous fresh fruits from the Southern Hemisphere in the previous decade. While apple exports had not grown as rapidly, they still increased by over 50 percent between 1974-76 and 1989. Codron argued that three main elements had contributed to that growth: the improvements in handling, storage and transportation, the entry of multinational trading companies into international trade in deciduous fruits and the exceptional dynamism of Chile. Chile's apple exports had risen more than sevenfold between 1974-76 and 1989.

The improvements in handling techniques, storage facilities, ports, containers, etc., and the increased speed of modern ships have also assisted the United States and France, who are major long-distance exporters. The entry of multinationals such as United Brands (of Chiquita banana fame), Del Monte, Polly Peck, and Castle and Cook (Dole) into production and marketing of apples in the Southern Hemisphere, has been noted on a relatively smaller scale in the Northern Hemisphere where many large operators were already firmly established and the multinationals face stiffer competition.

TABLE 10.1. World Production and Exports of Fresh Apples, Selected Years

Period	Production (1000 metric tons)	Exports (1000 metric tons)	Exports/Production (percent)
1948-52	13,512	519	3.8
1962-63	19,000	1,421	7.5
1969-71	28,309	2,180	7.7
1979-81	34,551	3,305	9.6
1986-88	40,114	3,621	9.0

SOURCE: UNFAO. *Production Yearbook* (annual).
UNFAO. *Trade Yearbook* (annual).

However, if demand for fresh fruits worldwide continues to grow, multinational companies can be expected to make strenuous efforts to increase their share of the business.

While technology has increased the volume and share of long-distance trade in fresh apples, the bulk of the trade continues to be between neighboring countries such as France and Britain or Italy and Germany. Reliable data on world apple trade flows is not readily available. For example, the same trade flow is often reported by both the exporting and importing country, but with differing figures. This can arise from differences in coverage, completeness and accuracy of reporting, delays in rerouting in transit, and other factors. Vidyashankara and Wilson (1989) developed a system for reconciling such problems, and reported estimates of apple trade flows between 76 exporting countries and 138 importing countries. However, since there was little or no trade between most countries, in their analysis of trade flows for the years 1962 through 1986 they looked at only 26 exporters and 30 importers.

Four of the top 10 exporters in 1986 were also among the top 10 importers, although it is likely that much of the figures for Belgium-Luxembourg and the Netherlands were pass-through trade (Table 10.2). For all the major Northern Hemisphere exporters and for Argentina, the major export destination was a close neighbor. In most cases, imports and exports were dominated by a single partner; for example, the United Kingdom was the major destination for French apple exports and France was the United Kingdom's major supplier. Only in the case of Southern Hemisphere exporters, Chile, New Zealand, and South Africa, were their primary markets at a long distance.

An analysis by this author of major trade flows for fresh apples in 1990-91 showed their continued strong regional concentration (Table 10.3). Almost half of all fresh apple trade moved within Europe, over 37 percent within the European Community. Western Europe (the EC and other) provided a market for a further one-quarter of world supplies from outside its borders. Most Southern Hemisphere exports went to Western Europe. Eastern Europe/USSR, and North America were their own primary suppliers. Only the Mideast and Other Asia depended primarily on suppliers from other regions. However, together they still took only one-tenth of world exports.

BARRIERS TO TRADE

The future growth of trade in fresh apples will depend heavily on the success of efforts to remove the many barriers that have persisted in agricultural trade. These barriers have normally been classified as either tariff or nontariff barriers. A tariff is essentially a tax imposed on an imported item either to raise government revenue or to raise the cost (and thus lower the demand) for the imported item. Tariffs are popular with governments and domestic consumers, less so with importers or consumers. Tariffs may be set as a fixed dollar amount per item or per unit imported (e.g., pounds, kilograms, bushels). They may also be set as a fixed percentage of the value of the item, called "ad valorem" tariffs.

Over time, tariffs have become less troublesome barriers to trade. If they are set as a fixed dollar amount, inflation reduces their relative impact. If they are set as a fixed percentage, they are relatively transparent and thus obvious targets for reduction in free-trade negotiations. As tariff rates have fallen, industry groups have pressed for, and governments have erected, more troublesome nontariff barriers.

In his pathbreaking study of nontariff agricultural trade barriers,

TABLE 10.2. Top Ten Exporters and Importers of Fresh Apples and Their Major Trading Partner, 1986

Rank	Exporter	Major Destination	Major Importer	Supplier
1	France	United Kingdom	West Germany	Italy
2	Chile	Netherlands	United Kingdom	France
3	Italy	West Germany	Netherlands	Chile
4	New Zealand	Belgium-Lux	USSR	Poland
5	Poland	USSR	Belgium-Lux	New Zealand
6	United States	Canada	United States	Canada
7	South Africa	United Kingdom	Austria	Hungary
8	Netherlands	West Germany	France	Netherlands
9	Argentina	Brazil	Canada	United States
10	Belgium-Lux	West Germany	Saudi Arabia	Chile

SOURCE: Vidyashankara, Satyanarayana, and Wesley W. Wilson. 1989. *World Trade in Apples: 1962-1987.* Pullman, WA: Washington State University, IMPACT Center, Information Series, No. 31.

Hillman (1978) argued that they had become "a major threat to both international cooperation and the future of efficient agriculture in the world." Hillman focused on thirteen major categories:

1. Quantitative restrictions such as quotas, embargoes or restrictive licensing.
2. Licensing requirements (licenses often being denied).
3. Variable levies, which may vary by formula or at the discretion of public officials.
4. Export subsidies.
5. Minimum import prices (usually set high to discourage imports).
6. Supplementary charges in addition to regular tariffs.
7. Import calendar, permitting imports only in certain months or varying levies by time of year.

8. Conditional imports.
9. State trading.
10. Mixing regulations.
11. Health and sanitary regulations.
12. Marketing standards and labelling.
13. Bilateral agreements.

For selected countries, Hillman noted that nine of the 13 categories were being openly applied to fruits and vegetables. In the 1980s, all 13 of these nontariff barriers could be found applied to trade in apples in some part of the world.

Barriers are occasionally even more prohibitive for processed products. For example, even within the European Community, many barriers to free movement of food and drink products persisted in the 1980s. A 1988 issue of Business International reported on a survey of 11,000 European businessmen concerning intra-Community trade barriers. Administrative barriers, frontier delays and costs, standards and technical regulations, implementation of Community law, differences in taxes, transport market regulations and capital market restrictions were all seen as barriers. These were targeted for elimination under the EC 1992 program for European integration. However, they would still be barriers to third-country traders.

While various trade barriers appear complex and difficult to evaluate, the effects are quite predictable. They either directly raise the price of imports or they reduce the supply, thus indirectly raising price. In contrast, export subsidies tend to lower the price of the subsidizing country, increase the supplies available and depress the price to all other suppliers. The major uncertainty about the effects of trade barriers arises when countries retaliate against each other, particularly with novel, erratic or deceptive barriers. In many cases, politicians or bureaucrats will approve a new nontariff barrier to buy short-term favor at home, reckoning (often correctly) that they will not get the blame for the inevitable retaliation.

GENERAL AGREEMENT ON TARIFFS AND TRADE

Governments have recognized that the only way in which a lid can be kept on protectionist tendencies and self-defeating retaliation

TABLE 10.3. Major Regional Trade Flows of Fresh Apples, Crop Year 1990-91 (percent)

Destination	European Community	Other West Europe	E. Europe /USSR	N. America	Southern Hemisphere	Mideast	Other Asia	Total
European Community	37.3	0.1	0.3	1.8	20.8	0.1	*	60.4
Other West Europe	1.7	0.1	1.9	0.6	1.9	*	*	6.2
E. Europe/USSR	0.4	*	7.2	*	*	*	*	7.6
N. America	*	*	*	4.4	1.3	*	*	5.7
C. America	*	*	*	0.5	0.6	*	*	1.2
S. Hemisphere	*	*	*	*	3.0	*	*	3.0
Mideast	*	*	*	0.8	2.2	1.9	*	4.9
Other Asia	*	*	*	4.4	0.6	*	0.2	5.3
All Other	2.7	*	0.2	0.8	1.9	0.3	*	5.8
TOTAL	42.1	0.2	9.5	13.4	32.2	2.3	0.2	100.0

SOURCE: USDA, FAS. Fresh Deciduous Fruit Reports (unpublished, additional occasional attaché reports).

* Less than 0.1 percent.

is by bilateral or multilateral agreements between countries. Only in this way can the principles of comparative advantage and the resultant specialization be allowed to operate to the benefit of all countries. Because of the protectionist excesses of the 1930s, there was widespread support after World War II for an international organization to help initiate and police freer trade. A proposal for an International Trade Organization (ITO) in 1948 was killed when the U.S. Congress failed to ratify it because it appeared to preempt some of the sovereign power of the United States. Eventually, a weaker organization established at the same time, the General Agreement on Tariffs and Trade, better known by its initials, GATT, became the vehicle for freeing international trade.

GATT maintains a small secretariat in Geneva and has little power to police or enforce agreements. It has operated by means of a series of negotiating rounds between member countries. During these rounds, members have successfully negotiated barriers downwards and have agreed on trading rules and complaint procedures. An important principle in spreading the benefits of concessions has been the most-favored-nation principle. Members agree to grant all other members the same concessions on any item as those granted to the most-favored trading partner. As GATT membership has grown to over 100 countries, this principle has spread concessions worldwide. Dispute resolution has been managed through a system where a complaint is filed, a panel of members reviews the case and a ruling is handed down. The loser in the complaint, usually reluctantly, and sometimes belatedly, complies because of the offsetting benefits of continued GATT membership.

The GATT system worked reasonably well between 1950 and 1970 when the emphasis was on reducing tariffs on industrial goods. However, as Hillman (1978) noted, by the early 1970s expansion of nontariff barriers was becoming a threat to the world economy. Nowhere was this more apparent than in agriculture. Hathaway (1987) succinctly explained how this had arisen. The GATT rules were set up in the immediate post-World War II era, when the U.S. was the dominant economic power. The General GATT rules were agreed upon by members, and governments brought their practices in line with these rules. "For agriculture, the process was exactly the reverse. The GATT rules were written to fit the agricultural programs then in existence, especially in the United States" p. 104). While GATT rules generally discouraged export subsidies, they were waived where they would countermand domestic farm programs. A similar exemption was permitted for quotas on agricultural products.

During the 1950s and 1960s, as the economies of Western Europe, East Asia and other developed countries expanded, and their agricultural sectors shrank, their protection for agriculture increased. As tariffs declined, governments became increasingly ingenious about the ways in which they assisted their agricultural sectors and attempted to disrupt agricultural imports.

By the mid-1980s, the farm programs of the U.S., the European Community and Japan together were costing their economies $300 billion a year. In the 1970s, U.S. exports of agricultural products had been growing rapidly and contributing significantly to the balance of payments. By the mid-1980s, the U.S. share of the world market had fallen dramatically, while the deficit in the balance of payments had climbed. Thus, when a new round of GATT negotiations got under way in Uruguay in 1986, the U.S. was in the lead in placing reform of agricultural trading rules on the agenda.

GATT had not been the only free trade initiative in the postwar era. A number of clusters of neighboring countries sought to speed the elimination of barriers to mutual trade by establishing regional free trade areas. Under GATT rules, these free trade areas could adopt a common external tariff for third countries. The most successful regional free trade experiment, the European Common Market (later to evolve into the European Community), while freeing up industrial markets, used its common agricultural policy to discriminate against agricultural trade from third countries. A second group of western European countries, not willing to open up their agricultural markets, formed the European Free Trade Association. Both the EC and EFTA added considerably to the free trade movement. Common markets for Latin American and Central American neighbors failed because member countries were unwilling to accept the painful adjustments. The centrally planned countries of Eastern Europe and the Soviet Union developed their own variant of a common market outside the aegis of GATT. They used limited specialization in production and trade on a barter basis. In their dealings with third countries, they generally used centralized state trading which was tacitly accepted by GATT if the countries were nondiscriminating and limited their price markups.

In the 1980s, the seventh GATT negotiating round, the Tokyo Round, made little headway in solving multilateral trade dilemmas. The U.S. General Accounting Office was asked by Congress to investigate whether the U.S. should make a last effort to reform GATT, should settle for its current ineffectiveness or should attempt to find an alternative system for promoting freer trade. The GAO report (1985) recommended that a multilateral system was preferable to a multitude of incompatible bilateral agreements.

At the same time that GATT was struggling, the European Community itself was in crisis. After a quarter century, trade within the Community continued to be hampered by numerous internal barriers. The Community was slipping behind North America and Asia in competitiveness. The Community's response was to set a target for removal of the major barriers and for completion of a single European market by the end of 1992. COMECON, the association of centrally planned countries, was also falling apart as the economic, social, and political fabric of their societies was collapsing, In 1989, the Berlin Wall fell. Within two years, all the countries of Eastern Europe had held elections, COMECON had evaporated, and the Union of Soviet Socialist Republics had been peacefully dissolved.

Thus, GATT negotiations, which had begun in 1986 in a world split between East and West in a pattern that had remained frozen for almost 40 years, found itself struggling for agreement in a world where many of the participants were watching the established order crumble while its replacement remained uncertain. As Germany poured its resources into its unification effort, the other European Community members scrambled to retain German interest in the single European process. Third countries, fearing that the European Community would become a protectionist "Fortress Europe," sought to lessen the impact. The European countries of EFTA negotiated a treaty with the EC on a European Economic Area which would include 18 Western European countries. A number of EFTA members, including Austria, Switzerland, and Sweden applied formally for EC membership. The newly democratized countries of

Eastern Europe set themselves the target of making their economies competitive enough to gain membership in GATT, the International Monetary Fund, other international agencies, and eventually full membership in the European Community.

The United States and Canada entered into a free trade agreement in 1989. Canada feared being excluded from the U.S. market as protectionist sentiment ebbed and flowed. The U.S. sought the Canada-U.S. Free Trade Agreement (CUSTA) both as a model for GATT and as a warning to its GATT partners that North America, too, could retreat into protectionism unless a satisfactory GATT agreement was reached. CUSTA contained the type of provisions for reducing of subsidies and freeing of trade in agriculture that the U.S. had been pressing for in GATT. Subsequently, in 1992, the U.S., Canada, and Mexico agreed to a similar North American Free Trade Agreement. Australia and New Zealand also established a free trade area and additionally reformed their economies in anticipation of competing without subsidies in a world economy.

All over the world, many other countries began to lower protective barriers and reduce government intervention in their economies in order to gain the expected benefits of freer trade. Mexico, Argentina, Nigeria, Indonesia, the Philippines, India, and Pakistan were among the countries with a long tradition of protectionism who began to open up their economies.

Ironically, during this same period, the GATT negotiations had reached an impasse. One of the major sticking points was agricultural subsidies. As part of any final deal, the U.S. and other major agricultural exporters were insistent that the European Community must make "substantial, progressive reductions" in trade-distorting subsidies on agricultural products. However, faced with depressed prices and the threat of even more agricultural imports from Eastern Europe, the Community could not hold the support of its members for a package that the U.S. or its allies would accept.

NONTARIFF BARRIERS

Without that breakthrough on subsidies, the Uruguay Round could not get agreement on some of the other major problems which continued to impede trade in agricultural products such as quantitative restrictions, state trading or voluntary restraint agreements. In the case of fresh produce items such as apples, phytosanitary restrictions had become a particularly insidious nontariff barrier. Tied in with these restraints were the need for new rule-making processes, dispute settlement mechanisms and monitoring and enforcement of trading rules within GATT.

However, any GATT agreement would be the beginning, not the end, of an attack on trade barriers affecting agriculture. For example, in the month that Canada and the United States initiated a new free trade area, the Canadian International Trade Tribunal found the United States guilty of dumping Delicious apples in the Canadian market at below cost and effectively imposed a minimum import price by setting "normal" values below which these apples could not be imported. This decision was taken despite the fact that in almost every year, in every fresh apple market around the world, the less popular grades and sizes have to be sold below the average cost at shipping point.

In many produce items, fluctuations in supply mean that in at least two years out of five, the average return at shipping point is below the average cost at shipping point. The more durable growers survive low-price years if they can recover at least their marginal costs. They use up their capital reserves, use off-farm income, or cut into their family living expenses to stay afloat. This sort of optimism that next year will be better is what enables farmers (and apple growers) to stay in the business despite heavy losses in any particular year. Even though Canadian growers operate on that same principle in both the domestic and export market, the Canadian International Trade Tribunal was willing to take punitive measures against U.S. growers for following normal sales practices.

In 1991, following continuing economic difficulties, Canadian apple growers were lobbying the Canadian federal government to set up a supply management agency which would have the power to limit or prohibit imports. This would have contravened the Canada-U.S. free trade agreement, would have been very questionable under existing GATT rules and would have been contrary to the final agricultural GATT proposals of the Uruguay Round.

The European Community had imposed "voluntary" restraint

agreements on Southern Hemisphere shipments of fresh apples in the mid-1980s. The EC authorities hoped that late storage apples from EC fall harvests could replace some of the Southern Hemisphere Apples harvested in the spring. However, consumers in the Community were willing to pay more for the reduced supply of Southern Hemisphere apples rather than buy poorer quality EC apples out of storage. Despite this failure, the EC persisted in its voluntary restraints and even turned back Southern Hemisphere shipments in transit in 1988.

The United States also rejected apples from certain suppliers for various reasons. Imports from France were rejected because of the presence of an insect, the pear leaf blister moth, not found in the United States. South African imports were suspended between 1987 and 1991 as a protest against apartheid in that country. Many other countries made similar protests against apartheid. However, the United States was one of the champions in finding causes for trade restrictions of one kind or another. Human rights, Arab boycott prohibitions, demands for emigration reform, protests against countries deemed to be particularly unfair traders, and condemnations of regimes considered "undesirable" in various ways were used as rationales for trade sanctions. While few of these sanctions directly affected apples, they were a constant reminder to trading partners of the unpredictability of U.S. trade relations.

Japan held a special place in the apple trade picture for the predictability of its barriers. Spurred on by the political power of its 95,000 small fruit growers, the Japanese authorities performed a classic stonewalling operation to keep out all fresh apple imports. The story begins in 1970, when the U.S. sought access for its sweet cherries to the Japanese market. These had been prohibited by the Japanese because they were grown in orchard areas where codling moth is endemic.

Arguments of U.S. experts that codling moths did not infest sweet cherries did not persuade the Japanese authorities. Eventually, a research project was initiated in which codling moths were placed in sealed plastic tents around sweet cherry trees, forced by hunger to attack the sweet cherries, and then the harvested cherries were fumigated to demonstrate that all codling moth life stages were eliminated. After seven years of patient research and tedious negotiations, in 1978 Japan permitted fumigated U.S. sweet cherries to enter the Japanese market during a limited window that would not interfere with domestic Japanese sweet cherry supplies.

In 1980, efforts were begun to develop a similar protocol to allow U.S. fresh apples into Japan. Despite the proven effectiveness of the fumigation technique, the Japanese authorities found numerous pretexts for demanding additional assurances. When it appeared that negotiated protocols on codling moths were acceptable, the Japanese raised additional concerns about fire blight (normally a problem in pear trees) and the lesser apple worm (virtually unknown in the Pacific Northwest exporting region). After over a decade of efforts to meet the requirements of the Japanese by jumping through the scientific hoops, it was clear to many in the U.S. apple industry that science could not provide answers in situations where the adequacy of scientific results was weighed by the country with a vested interest in rejecting those results.

Equally tricky issues were raised by trade barriers erected in the name of consumer protection. For example, Germany prohibited the use of edible waxes on apples, at the same time that other countries saw zero health problems with such waxes. Without an international standards organization to adjudicate such disputes, it will be difficult to separate genuine consumer hazards from pretense. Even more difficult are the issues where a majority of consumers in a country are so concerned about a substance such as Alar that elected officials feel obligated to ban that substance for political reasons. To what extent should an international body such as GATT insist that governments desist from such arbitrary actions without scientific evidence?

Without the presence of GATT or some other negotiating body, it may be very difficult to evaluate the legitimacy of phytosanitary complaints. For example, Mexican border officials have blocked entry of fresh fruits from the United States on grounds not used by countries such as Sweden or France which are recognized to have more effective phytosanitary controls. In contrast, Chile, which is generally recognized to be free of many fruit pests found in Argentina, is concerned that the transit of Argentine fruit may cause an infection which will lead to an embargo of Chilean fruit in third countries. In this case, the fate of Chile's very large export market could be damaged by an accidental infection from the other side of the Andes mountains. In such a situation, would an embargo by Chile on movement of Argentinean fruit through Chilean ports be a justified precaution or an unjustified trade barrier? Should the decision be one that an individual country like Chile could take unilaterally or should this be decided only by an international tribunal?

While the GATT negotiations were going on, the United States, through its Office of the Special Trade Representative, was much more aggressive in bilateral talks to limit various trade barriers in individual countries. The United States carried a number of bargaining chips into these negotiations, including the importance to other countries of continued access to the U.S. market for their goods and their need for favorable treatment from U.S. capital, technology and military and civilian assistance.

In particular, during the debt crisis and economic setbacks of the late 1970s and early 1980s, many countries had attempted to reduce their foreign debt by applying various curbs to "non-essential" imports. "Non-essential" was frequently defined to include imported fruits, including apples. Outright bans, quotas, restrictive licenses, advance deposits, limited import periods and other devices were used to discourage imports. The Northwest Horticultural Council, which represented the major U.S. exporters of deciduous fruits, led the U.S. effort from the industry side.

The joint efforts of the U.S. apple industry, the Office of the Special Trade Representative and other supporting public agencies led to the partial opening of markets for U.S. apples and other deciduous fruits in Mexico, Indonesia, Venezuela, the Philippines, Thailand, and other minor markets. In general, these partial openings were being gradually expanded for U.S. products, and the same concessions were being applied to comparable products from other suppliers. Thus, the U.S. effort was having multilateral effects in freeing trade.

The opposite was occurring in Northeast Asia where Japan's refusal to accept a codling moth control protocol was being used by South Korea to justify its closed market and by Taiwan, whose market had been open for more than a decade, to justify new economic and phytosanitary restrictions on imports. As long as Japan maintained its ban on U.S. apple imports, other suppliers would also be kept out of the Japanese market and South Korea and Taiwan would take no further steps towards liberalization of the fresh apple trade.

The U.S. has also taken a number of cases involving apples through the GATT complaint process (Northwest Horticultural Council, 1990). In 1989, a GATT panel found that the Norwegian opening date system was inconsistent with the GATT. Sweden had already phased out its opening date system, and Finland announced that it, too, would end its opening date system. In 1988, the European Community adopted a system requiring import licenses and a deposit for all apples imported into the Community between February and August 31, 1988. The U.S. requested a GATT arbitration panel which found that the Community's licensing system contravened the GATT.

MONOPOLY BOARDS IN APPLES

A further major unresolved issue, not covered by GATT, is whether or not monopoly boards for apples are anticompetitive. In many countries, a Ministry of Supply has monopoly control over imports of food items including apples. Clearly, under this sort of system, the decisions on the volume, quality or value of apples to import reflects not the free play of consumer demand, but the arbitrary allocation of government buying power according to political or administrative considerations, or the internal concerns of the ministry officials. In general, items such as fresh apples receive low priority in such a system.

A number of countries also give a single agency control of all exports of apples or apple products. Countries such as New Zealand and South Africa market worldwide under a single agency which decides which apples will be exported and at what price. The export agency controls storage for export, timing of shipments, transportation to be used, warehousing at destination, agents in the importing country, pricing, promotion and customer service.

The major objection of competitors to these monopoly export agencies, such as South Africa's Unifruco or the New Zealand Apple and Pear Marketing Board, is that either open or surreptitious government subsidies may give them an unfair competitive advantage relative to private traders. Countries whose currency was not freely convertible could provide the state export agency with favorable exchange rates that provided a price advantage in international markets. The major objections within country to these monopoly export boards were that private traders were denied access to an activity in which they could legitimately operate, and that monopoly led to waste, inefficiency and reduced return to growers.

Of course, the monopoly boards argue that allowing a number of traders to market a country's exports leads to lower quality standards, duplication of effort, less total marketing and promotional effort and other inefficiencies. To some extent, evaluation of monopoly boards is rooted in one's philosophy about when free enterprise is best, and when or to what extent that freedom should be curbed for the common good. The European Community has required new members to give up state-sponsored or state-owned monopoly boards, but has permitted them to be replaced by cooperative monopolies, where member firms voluntarily cede some or all of their international marketing activities to the national cooperative export board. It seems likely that such voluntary monopolies will continue to be legal under GATT.

The New Zealand Apple and Pear Marketing Board retained its monopoly control of both exports, imports and domestic marketing of apples and pears despite the drive to privatize most of the remainder of the New Zealand economy and despite the Australia-New Zealand free trade agreement. The Board has been accused of using that authority to subsidize exports of less popular varieties at the expense of more profitable varieties, to restrict imports unduly, and in giving preference to the export market over domestic consumer needs. The Board responded to these charges in 1990. In 1989, all apples exported received the same FOB price of $NZ 0.89 per kilogram, except for a two-cent premium for Gala and a one-cent premium for Royal Gala apples. In 1990, separate payment pools were instituted for different varieties with FOB prices ranging from $NZ 0.85 to $NZ 2.04 per kilogram.

The U.S. continued to complain that the Board discriminated against imports even in the Southern Hemisphere off-season, while large growing and packing entities in New Zealand continued to seek authority to compete in both the domestic and export markets.

Further expansion of multinational presence in the New Zealand
fruit industry is likely to keep challenges to the Board's monopoly
simmering. As the international trading system becomes less tolerant of government intervention in exports and imports, boards with
broad powers such as those enjoyed by the New Zealand Apple and
Pear Board will face future curtailment of those powers.

Chapter 11

Future Issues

The apple industry entered the twentieth century with a structure little different from that of biblical times. However, as it enters the twenty-first century it will be an industry subject to constant, rapid change.

Driving the whole process will be the need to be responsive to changing consumer tastes around the world. No longer will it be possible for growers to choose varieties that are easy to establish and maintain, or to keep the trees in the ground for half a century on the assumption that consumers will buy the apples from there. The consumer market is going to become increasingly segmented, with specific niches requiring different products. Many of those niches will be fickle, sensitive to temporary fads. They will require the industry to be much more nimble in its response than in the past.

The two major factors driving consumer markets are people and purchasing power. The world's population grew from 2.6 billion in 1950 to 3.7 billion in 1970 and 5.3 billion in 1990. It is expected to reach 7.2 billion by 2010, and 9.1 billion by 2030 (Urban and Trueblood, 1990). Thus, in 2030, total world population will be almost twice what it was in 1990. However, most of the growth in world population between 1990 and 2030 (perhaps 95 percent) will occur in the lesser-developed countries. Together, the United States, Canada, the European Community and Japan will contribute less than two percent of population growth. By 2030, the population of many of the richer countries will be either static or falling.

The age distribution of the population is also changing in predictable ways. In general, as countries experience economic growth and become more urban and industrialized, the birth rate drops dramatically. At a certain level of affluence, the birth rate is barely sufficient to maintain a stable population. However, affluence also brings improvements in medical care and in the average longevity of a population. The average age and the proportion of the population in older age groups rise. This process is more advanced in the United States, Japan and Western Europe. In the twenty-first century, countries like Korea, Taiwan and Singapore will also experience the phenomenon. In these countries, more of the purchasing power and of the assets will be in the hands of older consumers who are accustomed to having a wide range of choices, and who are discriminating in what characteristics they want from specific products.

Price will continue to be an important factor in affecting demand. However, price is a more critical factor when the product is relatively expensive, incomes are low, and product offerings are relatively homogeneous. For example, in buying an ounce of gold, price may be the only factor considered. In contrast, in most developed countries, a pound of apples may sell for one or two dollars, less than 10 minutes work at the average hourly wage. There will be many varieties of apples available and hundreds of possible alternative fruits, vegetables or snacks. Other factors such as aesthetics, the eating occasion, taste and texture, convenience, and the consumer's expectations will become more important than price in influencing demand.

As well as the momentary enjoyment, consumers in the 1990s will continue to be influenced by their perception of the cumulative effects of eating different foods. Mothers with infants and children worry about the effect of each food on the current health of the child and on the development of healthy bones, teeth, and skin. They worry about the adverse effects of pre- or postharvest chemicals or waxes. As their children reach the teen years, mothers become concerned about fat, salt, sugar, and cholesterol. Young adults in increasing numbers adopt those same concerns for their own health, worrying in particular about the effect of food on their appearance, attractiveness to the opposite sex, and vitality. These same concerns are moving forward into middle age and beyond, supplemented by the desire to avoid the problems of high blood pressure, clogged arteries, cancer, and other illnesses that have been linked to dietary habits. Even in retirement years, concern about the quality of life is leading to increasing pickiness about the foods eaten.

Undoubtedly, during the 1990s, the links between food habits and health will become clearer. Improved information will make it easier for consumers to weigh the risks of a particular food (from naturally occurring and applied chemicals) versus the benefits of that food to health and nutrition. Hopefully, for example, in some future Alar incident, data will be available on the relative benefits and risks to both the mass of consumers and to special groups such as the very young, the very old, pregnant or lactating women, or the chronically ill. However, many conflicting claims will remain. Some consumers will continue to express their concerns by calling for foods grown by organic methods, or with restricted chemical use, or with enhanced nutritional properties. Other consumers will remain indifferent or antagonistic to such concerns. Media events will continue to sway the proportion who are concerned, indifferent or unconcerned at any one time. Apple producers, processors and marketers will have to be adaptable to those changing market segments.

In the long-term, this will require the fresh apple industry to pursue a deliberate policy of product differentiation similar to that followed for decades by branded, packaged goods. Such differentiation rests on continuing market research to identify current and emerging consumer interests and concerns. Its execution requires positioning products in key market segments both for defensive and offensive purposes. In the apple juice markets, processors have developed juices of various strengths (from 100 percent fruit juice to fruit drinks that are primarily water), in various forms (fresh, chilled, frozen and canned), of different appearance (clear or cloudy), and so on. Thus, if consumer tastes swing toward a cloudy, frozen juice drink, the processor has an acceptable product either ready for or in the marketplace. If, through heavy promotion, a major competitor threatens to expand the canned, clear segment, the processor can quickly promote an alternative entrant in that category.

Such a capability is likely to emerge in the fresh apple industry in one or two ways. National boards may begin to emulate the New Zealand Apple and Pear board which has developed a stable of products capable of tapping both traditional and new markets in many different countries. In addition, it is investing in the breeding and product-testing programs needed to bring along new varieties to replace those that are losing favor with consumers. An alternative way may be the more aggressive expansion of large multiproduct companies into product differentiation in apples. While these companies clearly have the financial strength and marketing muscle, they are reluctant to invest heavily in new product development unless they can have some proprietary claim on the results. They rarely have the in-house capability in research and development in tree fruits. That normally is found only in publicly funded research centers run by governments or universities. However, if the process of developing and testing new apple varieties could be speeded up or reduced in cost through biotechnology or other techniques, private companies would be more likely to initiate product differentiation programs in fresh apples.

In other products, both food and nonfood, public-private partnerships have been forged to help strengthen a national effort. However, the different goals of the partners make these efforts difficult to hold together. In addition, many of the larger produce corporations are multinational in scope, so there is the possibility of conflict between their broad corporate goals and narrow national interests.

While the control of product development in fresh apples is uncertain, there is little doubt about the growing importance of multiproduct corporations in fresh apple marketing in general. Economies of scale in fresh packing and processing have increased the capital needs of the industry and made ownership of apple packing and processing plants more attractive to corporate investors. However, the concentration of ownership and control of all food retailing and wholesaling has also made it attractive for corporations with a national sales force to handle many fruit products, including imported items like bananas and pineapples and domestic items like apples and pears. Such multiproduct companies no longer expand plant capacity in response to availability of local raw materials, whether apples or pineapples. Rather, they expand or contract their apple division in response to the profitability or return on investment to be gained from apples.

These multiproduct companies tend to be more sensitive to consumer and customer needs. They have extensive experience in marketing research and the capability, where needed, of employing sophisticated (and expensive) consultants. They can also gain economies of scale in integrated information processing. For example, their computers can be made interactive with those of major wholesalers and retailers. Order-taking, shipping instructions, inventory management, accounts receivable and accounts payable, and other management information can be streamlined.

Companies dependent on a single product or on a number of products from a single location will face a strategic dilemma. They can retain the role of a specialized supplier. If they choose to remain small, they will have to be capable of meeting some specialized niche that justifies a large retailer retaining them as a supplier. If they remain specialized, but grow larger to gain economies, they will become more vulnerable to a downturn in the apple economy than their multiproduct rivals. Alternatively, they can themselves become multiproduct, multilocation producers. Juice processors have been able to diversify their product line rather easily, because raw materials from all over the world are readily available in concentrate form. Most packers of fresh apples would have to build operations in new locations if they were to diversify into grapes or oranges, or into new continents if they wished to add bananas or pineapples. In the latter case particularly, they would face a steep learning curve relative to established multinationals.

Of course, diversification across hemispheres has a particular marketing implication. For example, much of the growth of the western U.S. apple industry in the 1960-80 period resulted from a gradual extension of effective storage life until selected fruit could be marketed for 12 months after harvest. However, the expansion of apple exports from the Southern Hemisphere means that there continues to be an increasing supply of new season apples available in the Northern Hemisphere beginning February and March of each year. Producers with operations in both hemispheres can be assured of always having the freshest available apples. Late season apples from Northern Hemisphere storage will continue to face stiff competition from fresher Southern Hemisphere products.

The flow of trade will, of course, be influenced tremendously by the general reaction of nations to the opposing forces of liberalization and protectionism. Since the 1980s, many nations have attempted to reduce their reliance on central planning, state trading and other restrictions on free market operations. However, free markets bring greater risks and greater possibilities of loss of capital. Only time will tell whether these societies can tolerate the painful transition to a market economy.

Similar uncertainty overhangs the many bilateral and multilateral initiatives to promote free trade. The General Agreement on Tariffs and Trade (GATT) has been struggling against many odds to increase the transparency and fairness of the international trading system while dismantling old barriers and attempting to thwart new barriers.

Regional trade groupings such as the European Community, and free trade areas in North America, Australasia, Latin America, and Asia offer possibilities for enhancing the movement towards freer trade or for becoming protectionist blocs. Trade in apples and apple products will be influenced by the result of that struggle. It is difficult to predict how rapidly or how effectively trade in fresh fruit will be freed.

Much will depend on the health of economies around the world. Economic growth generally stimulates employment. Societies are less defensive of the sectors that are hurt by imports if other sectors of their economy are growing. In turn, economic growth usually translates into higher personal incomes. As incomes rise, consumers are more likely to demand the quality and diversity of product that is only available through trade. Thus, the general tone of the future world apple market will be set by developments in ideologies, politics, trade negotiations, and economic conditions.

World apple supply will continue to respond, both to these market conditions and to natural, commercial and societal conditions. The land area that can be profitably planted to apples will continue to be sensitive to market price prospects, but will be less bound by traditional locations, technologies and growing practices. Irrigation technology has increased many times the acreage that can be planted to apples. Breeding programs and adaptive research are demonstrating opportunities for different varieties in numerous microclimates once considered unsuitable for apples. Advances in biotechnology will expand the range of opportunities further.

Governments will affect supply in numerous ways. While massive subsidies to large plantations from national or multinational agencies are less in vogue than in the 1970s and 1980s, such subsidization may continue in different guises in the 1990s. Aids to regional development, financing of fruit cooperatives, liberal bank credit, subsidized inputs such as water, power or energy, marketing concessions, tax relief, consumer subsidies, and numerous other schemes can be used to reduce the real cost of developing or operating orchards. Such policies tend to stimulate additional production that would not otherwise occur.

In pursuit of other social objectives, however, governments will often simultaneously impose costs on the apple industry which will discourage increased production. Governments will be under extensive pressure in the 1990s to discourage consumption of fossil fuels, agricultural chemicals, food additives, and other conventional inputs that have contributed in the past to higher yields and lower unit costs. Governments will require additional safeguards in the orchard, workshop, and processing plant; added controls of product health, safety, nutrition, waste products, and labelling; more protection of workers from medical costs, unemployment, disability or retirement; special facilities and services for special classes of workers such as the disadvantaged, racial or other minorities, the young, the elderly, etc.; and the burden of financing, record-keeping and self-monitoring of each required program.

Where the burden of such programs falls heavier on fruit growing than on other business activities, the effect will be to stimulate marginal shifts out of fruit growing. Or, if fruit growing in a more-developed country is impacted differently from fruit growing in a less-developed country, the effect may be marginal relocation of apple growing.

Until now, the apple industry approach to government regulation has been largely reactionary, seeking to delay or modify new rulings, and, if this fails, reluctantly adapting. The industry needs to join with the rest of society in reexamining (1) the desirability of the goals of these proliferating regulations, (2) the effectiveness of the programs in achieving the goals, and (3) the magnitude of the unforeseen or unwanted side-effects. In particular, it is important to identify whether or not different programs promote contradictory or offsetting goals, and the international ramifications of regulations imposed in one country on the business of another.

This will require the roping of many golden calves and the upending of many sacred cows. For example, a whole culture has evolved around the concept of "organic" produce; that is, produce grown only with the use of organic compounds. However, there is no evidence that the replacement of inorganic chemicals by organic chemicals will reduce the dosage of toxins entering the environment or the food supply. There is no evidence that human health may be benefitted. Indeed, in the short-run, higher prices of organic produce may lower the desirable intake of fresh fruits and vegetables. This is not to say that synthetic chemicals should not be carefully evaluated for their potential hazard to workers, the environment, and consumers. Rather, all chemical and nonchemical food production and preservation systems need to be objectively evaluated, not routinely praised or condemned based on emotionally charged labels.

Rationality in food import laws will also become an issue in the international arena. Some objective measures for assessing risk will have to be developed, some international standards of food safety established and some supranational panel empowered to adjudicate on the merits of competing claims.

Rationality will also become critical in maintaining a balance between new technology that will become available to the apple industry and the economic, social and environmental results of that technology. For example, the twentieth century has seen the capability for apple production extended from standard trees to dwarf trees to bushes to meadow plantings that can be mowed down after each harvest. In the next few decades we may see apples and apple trees altered in various ways through biotechnology. Apples may be grown in a test tube, without a tree. Such developments will further alter the ownership patterns and capital needs of the industry.

The technology widely employed in the apple industry in the 1990s will also come under increasing challenge for its monoculturism and its pervasive use of chemicals. Battles will rage around the application of acronyms such as IPM (Integrated Pest Management) and LISA (Low-Input Sustainable Agriculture) or terms such as sustainable or organic. Sustainability will be a particular issue because it encompasses not just the operations or product of an orchard that the word organic implies, but also the future fate of air, soil, water, and mankind. Postharvest treatment of apples with chemicals to maintain quality has also become increasingly controversial. Consumers will eventually have to decide whether the risks of chemical residues are offset by the aesthetic, nutritive, and enjoyment benefits of having fresh apples 12 months a year.

Technology will also play a major role in the development of the apple packing, processing and distribution sector. If current trends persist, automation, electronics, computers and other capital-intensive or labor-saving devices will become universal in the surviving apple packing plants, which will tend to become larger. These larger plants will be better able to provide a full line of fresh apple products to the more concentrated marketers and distributors. Electronic data interchange will, in turn, favor increased business between larger buyers and sellers and their transportation agencies.

The apple processing industry will also be changed by advances in technology now in the development stages. Methods for preserving the taste, flavor or odor of fresh apples in the processed product will be improved. Other goals such as increased speed of production, reduced energy use, and less waste will draw forth new processes. Very few of these new technologies are likely to favor small plants relative to large. As in most other facets of the apple industry, the small operator will survive only by serving specialty niches requiring unique effort or expertise.

The globalization of ideas, finance, markets and organizations will accelerate. Communication by satellite, radio, telephone, fax and variants yet unknown will affect every apple business. Decisions will be possible between sites more and more remote from each other. A buyer in London can already place a conference call connecting Canberra, Santiago, Pretoria, and Yakima to assess the current storage volume, quality and offer price of a given type of apple. An investor in Boston can weigh the costs and returns of an orchard planting in Brazil or Turkey. Most governments around the world are aggressively seeking to bring modern communications to every corner of their territory. The key to prosperity in a world where ideas, capital, enterprises and goods can move rapidly in any direction will be information.

It will be critical for participants in the apple business to acquire the information needed to survive in the global economy. As in all industries, apple producers, processors and marketers feel most comfortable acquiring more information on the part of the business they already know best: producers on production, marketers on marketing and so on. However, in a world of rapid change, each firm must know and understand the factors that are imposing change on their suppliers and customers. Producers must understand consumers, retailers must understand shippers, transportation agencies must understand growers.

In addition, all aspects of the apple industry need to stay informed of the changes in the external environment which will inevitably break into their traditional ways of doing business. The earlier each firm or organization in the apple industry can detect the new ideas, social movements or technologies which may have implications for their business, the more likely they are to make far-sighted decisions in adapting to coming changes. Those firms and organizations will fare best that can systematically scan, select, and evaluate the massive flow of information that now bombards them and can incorporate that knowledge into effective business strategies for the future.

References

Agriculture Canada. 1985. "The Economic Potential for Concentrated Apple Juice Production in Canada." Ottawa, Ontario. Working Paper 10/85.

Allison, Lisa, and Donald Ricks. 1986. "Analysis of Prices for Juice Apples." East Lansing, MI: Department of Agricultural Economics, Michigan State University. Presentation to USITC Investigation No. TA-201-59.

Al Staffy, Tawhid M., and A. Desmond O'Rourke. 1984. *Impact of Quality on Marketing Margins: A Case Study for Washington Apples*. Pullman, WA: Department of Agricultural Economics, Washington State University, Scientific Paper No. 6811.

Ames, Bruce, and Lois Gold. 1990. "Too Many Rodent Carcinogens." *Science* 249 (August 31): 970-971.

Australian Bureau of Agricultural Economics. 1988. "Japanese Agricultural Policies." Australian Government Publishing Service, Canberra. Policy Monograph, No. 3.

Bagley, Robert D. 1977. "Potentials for Exports of Deciduous Fruits from the United States to Western Europe." Master's thesis, Department of Agricultural Economics, Washington State University.

Baumes, Harry S., Jr., and Roger K. Conway. 1985. *An Econometric Model of the U.S. Apple Market*. Washington, DC: USDA, ERS, Staff Report No. AGES850110 (June).

Ben-David, Shaul, and William G. Tomek. 1965. *Storing and Marketing New York State Apples, Based on Intraseasonal Demand Relationships*. Ithaca, NY: Cornell University Agricultural Experiment Station, Bulletin 1007.

Bressler, Raymond G., Jr., and Richard A. King. 1970. *Markets, Prices and Interregional Trade*. New York: John Wiley and Sons.

Bulatao, Rodolpho A., Edward Bos, Patience W. Stevens, and My T. Vu. 1990. *World Population Projections*. 1989-90 edition.

Baltimore: The Johns Hopkins University Press (Published for the World Bank).

Bureau of Labor Statistics. *Consumer Price Index*. Washington, DC: U.S. Department of Labor (monthly).

Burns, Alfred J., and Victor G. Edman. 1970. *Prices and Spreads for Apples, Grapefruit, Grapes, Lemons, and Oranges Sold Fresh in Selected Markets, 1962/63-1966/67*. Washington, DC: USDA, ERS, Marketing Research Report No. 888.

_____. 1975. *Prices and Spreads for Selected Fruits Sold Fresh in Major Markets, 1967/68-1973/74*. Washington, DC: USDA, ERS, Agricultural Economic Report No. 295 (August).

Burns, Alfred J., and Joseph C. Podany. 1975. *Prices and Spreads for Selected Fruits Sold Fresh in Major Markets, 1967-68 - 1973-74*. Washington, DC: USDA, ERS, Agricultural Economics Report No. 295.

Cain, Jarvis L., and Eugene T. Shawaryn. 1976. *Maryland Consumers View Fresh Apple Marketing*. College Park, MD: Agricultural Experiment Station, University of Maryland, MD 890.

Castaldi, Mark. 1988. *World Apple Survey: Production Distribution and Trade*. Ithaca, NY: Cooperative Extension, Cornell University, XB015.

Cavalieri, Ralph P., and Marvin T. Pitts. 1991. "Non-destructive Sensor for Apple Firmness." Wenatchee, WA: *Proceedings of the 7th Annual Washington Tree Fruit Postharvest Conference, 1991*. Wenatchee, pp. 55-56.

Central Statistics Office. 1989. *Household Budget Survey, 1987*. Volume I. "Detailed Results for All Households." Dublin, Ireland: Government Publications Sales Office.

Codron, Jean-Marie. 1990. "L'Hemisphere Sud et la mondialisation des echanges de fruits temperes." Paper presented at an E.E.C. seminar in Crete, Greece (November).

Consumers' Research. 1989. "Does Everything Cause Cancer?" *Consumers' Research Magazine* 7(5): 11-18.

Cummins, James N., and Herb S. Aldwinckle. 1991. "Rootstocks for the Modern Orchard." Wenatchee, WA: *Proceedings of the Washington State Horticultural Association, 86th Annual Meeting, 1990, Yakima, pp. 14-24.

Dickrell, Peter A., Herbert R. Hinman, and Paul J. Tvergyak. 1987.

1987 Estimated Cost of Producing Apples in the Wenatchee Area. Pullman, WA: Washington State University Cooperative Extension Bulletin 1472.

Dunham, Denis. 1990. *Food Costs Review, 1989.* Washington, DC: United States Department of Agriculture, Economic Research Service, Agricultural Economic Report No. 636.

Edman, Victor G. 1964. *Prices and Spreads for Fresh Fruits and Vegetables.* Washington, DC: USDA, MED, ERS, Statistical Bulletin No. 340 (February).

England, W.B. 1959. *Operating Results of Food Chains in 1958.* Boston, MA: Harvard Business School, Bureau of Business Research, Bulletin No. 156.

European Community Commission. 1982. "Mechanisms of the Common Organization of Agricultural Markets-Crops Products." *Green Europe Newsletter* (Brussels) No. 189.

––––––. 1983. "The Common Agricultural Policy and the Food Industry." *Green Europe Newsletter* (Brussels) No. 196.

––––––. 1984. "Fruit and Vegetables. Why Products are Withdrawn from the Market." *Green Europe Newsletter* (Brussels) No. 205.

––––––. 1989. "The Common Agricultural Policy and the 1990s." *European Documentation Series* (Luxembourg) 5/1989.

Fisher, D.V. 1966. *High-Density Orchard for British Columbia Conditions.* Summerland, BC: Research Branch, Canada Department of Agriculture.

Florida Department of Citrus. 1987. *Orange Juice Strategy Study.* Gainesville, FL: Market Research Report (April).

––––––. 1987. *Distribution and Use of Orange Juice in Public Schools.* Gainesville, FL: Market Research Report (August 28).

––––––. 1988. *Annual Foodservice Survey, 1986-87.* Gainesville, FL: Market Research Report (March 3).

––––––. 1991. *Florida Citrus Outlook, 1991-92 Season.* Gainesville, FL: Economic Research Department, Working paper series.

Frick, Lawrence F., and Ulrich C. Toensmeyer. 1977. *Delaware Household Consumption of Fresh Apples.* Newark, DE: Agricultural Experiment Station, University of Delaware, Bulletin 4211.

Frumkin, Paul. 1989. "Consumer Trends Outlook: The 1990s." *Restaurant Business.* Vol. 9, No. 5, (May 20): 160-161.

Gardiner, Walter H. 1990. "The EC's Agrimonetary System: Pressures for Reform." Washington, DC: USDA, ERS, *Western Europe Agriculture and Trade Report RS* 90-4 (November): 53- 60.

German, Gene, and Gerard Hawkes. 1983. *Operating Results of Food Chains, 1982-83*. Ithaca, NY: Cornell University, September.

Government Accounting Office. 1985. "Current Issues in U.S. Participation in the Multilateral Trading System." Washington, DC: *GAO/NSIAD*-85-118 (September).

Greig, W. Smith. 1971. *Location Advantages in Applesauce Processing in the U.S. with Some Implications for the Washington Apple Industry*. Pullman, WA: Washington Agricultural Experiment Station, Bulletin 753.

Greig, W. Smith, and Leroy L. Blakeslee. 1975. *Potential for Apple Juice Processing in the U.S. with Implications for Washington*. Pullman, WA: Washington State University College of Agriculture Research Center, Bulletin 808.

Greig, W. Smith, and A. Desmond O'Rourke. 1972. *Apple Packing Costs in Washington, 1971*. Pullman, WA: Washington Agricultural Experiment Station Bulletin 755.

Gruen, Nina. 1989. "The Retail Battleground: Solution for Today's Shifting Marketplace." *Journal of Property Management* (40-43) July/August.

Handy, Charles R., and Daniel I. Padberg. 1971. "A Model of Competitive Behavior in Food Industries." *American Journal of Agricultural Economics* 53(2) (May): 182-190.

Hathaway, Dale E. 1987. *Agriculture and the GATT: Rewriting the Rules*. Washington, DC. Institute for International Economics, Policy Analyses in International Economics, No. 20 (September).

Heim, Manfred, and A. Desmond O'Rourke. 1988. *Overview of Japanese Fruit Consumption Trends with Emphasis on Apple Demand*. Pullman, WA: Washington State University IMPACT Center, Information Series No. 24.

Heller, Walter. 1990. "Supermarkets 2000: 45 'Insider' Predictions." *Progressive Grocer* (January): 24-30. Vol. 69, No. 1.

Heydon, Richard N. 1981. "The Potential for Washington Fruit

Exports to Selected Asian Countries." Master's thesis, College of Agriculture Research Center, Washington State University.

Heydon, Richard N., and A. Desmond O'Rourke. 1981. *Implications of Japanese Economic Growth for Imports of U.S. Agricultural Products with Special Reference to Deciduous Fruit Imports.* Pullman, WA: Washington State University College of Agriculture Research Center, Bulletin 0903.

———. 1982. *The Potential for Washington Fruit Exports in Southeast Asia, Excluding Japan.* Pullman, WA: Washington State University, Agricultural Research Center, XB0911.

Hiemstra, Stephen J., and Daniel B. DeLoach. 1962. *Growth Patterns in the Retail Grocery Business.* Davis, CA: California Agricultural Experiment Station, Bulletin 786.

Hillman, Jimmye S. 1978. *Nontariff Agricultural Trade Barriers.* Lincoln, NE: University of Nebraska Press.

Hinman, Herbert R., Paul Tvergyak, Brooke Peterson, and Marc Clements. 1992. *1992 Estimated Cost of Producing Red Delicious Apples in Central Washington.* Pullman, WA: Washington State University Cooperative Extension, Farm Business Management Report, EB 1720.

Ho, Foo-Shiung. 1974. "Computerization of Daily Decisions on Utilization of Receipts of Apples in a Single Product Apple Processing Plant." PhD diss. Department of Agricultural Economics, Virginia Polytechnic Institute.

Hyde, Gary M., Weihua Zhang, and Ralph P. Cavalieri. 1990. "Bruise Detection Using the Instrumented Sphere and a High Speed Video Camera." St. Joseph, MO. Paper presented at the 1990 International Winter Meeting of the American Society of Agricultural Engineers, Paper No. 90-1612.

Jussaume, Raymond A., Jr. 1991. *The Japanese Juice Industry: A General Overview.* Pullman, WA: IMPACT Center, Washington State University, Information Series No. 47.

Kaplan, A.D.H., J.B. Dirlam, and R.F. Lanzilotti. 1958. *Pricing in Big Business.* Washington, DC: The Brookings Institute Press.

Kaufman, Phillip R., and Charles R. Handy. 1989. *Supermarket Prices and Price Differences.* Washington, DC: USDA, ERS, Technical Bulletin No. 1776 (December).

Kelsey, Myron P. 1990. *Business Analysis Summary for Fruit*

Farms, 1988 Telfarm Data. East Lansing, MI: Michigan State University, Department of Agricultural Economics, Agricultural Economics Report No. 532.

King, Norm B., and A. Desmond O'Rourke. 1977. *The Economic Impact of Changes in Pesticide Use in Yakima Valley Orchards.* College of Agriculture Research Center, Washington State University, Bulletin No. 841.

Kortlave, Ing C. 1991. "Components of Successful Dutch Apple Production." Wenatchee, WA: *Proceedings of the Washington State Horticultural Association,* 86th Annual Meeting, 1990, Yakima, pp. 26-34.

Lee, Mei Li. 1986. "Cost Competitiveness of Apple Production in British Columbia versus Washington State." Vancouver, BC: Department of Agricultural Economics, University of British Columbia, Working Paper 18/86.

LePage, D.K., and F.H. Jackson. 1988. "Horticultural Establishment and Production Costs." Christchurch, New Zealand: Economic Consultancy Unit, MAFCorp, Ministry of Agriculture and Fisheries, Technical Paper 1/88.

Lusztig, Peter A. 1990. *Report of the Commission of Inquiry–British Columbia Tree Fruit Industry.* Vancouver, BC: Commission of Inquiry–British Columbia Tree Fruit Industry.

MacGregor, Joan M., and Robert L. Jack. 1979. *Profile of Fresh Apple Purchasers in West Virginia.* Morgantown, WV: Agricultural and Forestry Experiment Station, West Virginia University, Bulletin 672.

Marion, Bruce W., Willard F. Mueller, Ronald W. Cotterill, Frederick E. Geithman, and John R. Schmelzer. 1977. *The Profit and Price Performance of Leading Food Chains, 1970- 74.* Washington, DC: Joint Economic Committee, Congress of the United States.

McLaughlin, Edward W. 1983. "Buying and Selling Practices in the Fresh Fruit and Vegetable Distribution System: Implications for Vertical Coordination." PhD diss. Department of Agricultural Economics, Michigan State University.

_____. 1983. "The Fresh Fruit and Vegetable Marketing System: A Research Summary." East Lansing, MI: Department of Agricultural Economics, Michigan State University.

Ministry of Agriculture, Fisheries and Food. 1990. *Household Food Consumption and Expenditure, 1989*. London: Her Majesty's Stationery Office, Annual Report of the National Food Survey Committee.

National Commission on Food Marketing. 1966. *Organization and Competition in Food Retailing*. Washington, DC. Technical Study No. 7 (June).

National Restaurant Association. 1988. *Foodservice Industry 2000*. Washington, DC. Current Issues Report.

_____. 1989. *Foodservice Industry 2000*. Washington, DC. Current Issues Report.

_____. 1990. "1991 Foodservice Industry Forecasts." *Restaurant USA* 10 (11) (December): 13-36.

Northwest Horticultural Council. 1990. *Trade Barriers Report and Five Year Summary*. Yakima, WA (December 31).

Norton, Robert A., and Jacqueline King. 1987. *Apple Cultivars for Puget Sound*. Cooperative Extension, College of Agriculture and Home Economics, Washington State University, EB 1436.

OECD (Organization for Economic Co-Operation and Development). 1991. *The Apple Market in OECD Countries*. Paris, France.

O'Rourke, A. Desmond. 1974. *Factors Affecting Major Marketing Decisions for the Washington Apple Crop*. Pullman, WA: Washington State University College of Agriculture Research Center Bulletin 793.

_____. 1980. *The Role of Direct Marketing in Washington Agriculture*. Pullman, WA: Washington State University, College of Agriculture Research Center, Circular Number 0890.

_____. 1986a. Econometric Model of the United States Apple Juice Industry. Pullman, WA: Presentation to US, ITC. Investigation No. TA-201-59. Washington State University College of Agriculture and Home Economics, Agricultural Research Center.

_____. 1986b. *Outlook for Granny Smith Apples in Washington State*. Washington State University College of Agriculture and Home Economics, Agricultural Research Center. Research Bulletin 0971.

_____. 1990. "Anatomy of a Disaster." *Agribusiness* 6(3): 417-424.

Paarlberg, Don. 1980. *Farm and Food Policy.* Lincoln, NE: University of Nebraska Press.

Parker, Russell C. 1976. "The Status of Competition in the Food Manufacturing and Food Retailing Industries." Madison, WI. NC Project 117 Working Paper Series W8-6 (August).

Pasour, E. C., Jr. 1990. *Agriculture and the State.* New York: The Independent Institute and Holmes and Meier.

Preston, Lee R. 1963. *Profile, Competition and Rules of Thumb in Retail Food Pricing.* Berkeley, CA: University of California, Berkeley, Institute of Business and Economic Research.

Price, David W. 1973. *Intraseasonal Demand, Supply, and Allocation of Washington Apples.* Pullman, WA: Washington Agricultural Experiment Station, Technical Bulletin 75.

Produce Marketing Almanac. 1981. Newark, DE: Produce Marketing Association, Inc.

The editors of *Progressive Grocer.* "Product Performance in the Super Store." Vol. 54, No. 7. (July): 36-125.

Putnam, Judith Jones. 1989. *Food Consumption, Prices and Expenditures, 1967-88.* Washington, DC: U.S. Department of Agriculture, Economics Research Service, Statistical Bulletin No. 804.

Putnam, Judith Jones, and Jane E. Allshouse. 1991. *Food Consumption, Prices, Expenditures, 1968-89.* Washington, DC: U.S. Department of Agriculture, Economic Research Service, Statistical Bulletin No. 825.

Ritson, Christopher, and Alan Swinbank. 1984. "Impact of Reference Prices on the Marketing of Fruit and Vegetables." Newcastle upon Tyne, England. *Price and Market Policies in European Agriculture, Proceedings of the Sixth Symposium of the European Association of Agricultural Economists*, pp. 248-266.

Roucaud, M. Y., and B. Yon. 1983. "Managing the Development of Agrofood Industries in Rural Areas." Paris, France. *OECD Food Industries in the 1980s*, pp. 73-83. OECD.

Schotzko, R. Thomas. 1981. *Apple Packing Systems: Comparison of Selected Costs Between Conventional and Presize Systems.* Pullman, WA: College of Agriculture, Washington State University, Extension Bulletin 0935.

Scott, Michael. 1971. "The Value of a Processing Diversion to the

Washington Apple Industry." Master's thesis, Department of Agricultural Economics, Washington State University.

Smock, R. M. 1949. *Controlled Atmosphere Storage of Apples*. Ithaca, NY: Cornell Extension Bulletin 759.

Stanton, B. F., and B. A. Dominick, Jr. 1964. *Management and Cost Control in Producing Apples for Fresh Market*. Ithaca, NY: Cornell University Agricultural Experiment Station, Bulletin 1001.

Tomek, William G. 1968. *Apples in the United States: Farm Prices and Uses, 1947-75*. Ithaca, NY: Cornell University Agricultural Experiment Station, Bulletin 1022.

Tomek, William G., and Kenneth L. Robinson. 1972. *Agricultural Product Prices*. Ithaca, NY: Cornell University Press.

Trotter, C. E., and T. A. Brewer. 1977. *Consumer Reactions to Fresh Apples Marketed in Allentown-Bethlehem, Pennsylvania, 1974-75*. University Park, PA: Agricultural Experiment Station, Pennsylvania State University, Bulletin 816.

Tweeten, Luther. 1991. *The Economics of an Environmentally Sound Agriculture (ESA)*. Columbus, OH: Department of Agricultural Economics and Rural Sociology, Ohio State University, ESO, 1784.

United Nations Food and Agriculture Organization. *Production Yearbook*, annual, selected issues.

United Nations Food and Agriculture Organization. *Trade Yearbook*, annual, miscellaneous issues.

U.S. Department of Agriculture. 1957. *Farm-Retail Spreads for Food Products*. Washington, DC: Agricultural Marketing Service, MRD, Miscellaneous Publication No. 741 (November).

_____. 1965a. *Foreign Economic Growth and Market Potentials for U.S. Agricultural Products*. Washington, DC: ERS, Development and Trade Analysis Division, Foreign Ag. Econ. Report 24.

_____. 1965b. *U.S. Food Consumption, Sources of Data and Trends, 1909-63*. Washington, DC: Economic Research Service, Statistical Bulletin 364.

_____. 1966. *Homemakers' Use of and Opinions About Selected Fruits and Fruit Products*. Washington, DC: Statistical Reporting Service, MRR 765.

_____. 1977a. *Cost Components of Farm-Retail Spreads*. Washington, DC: ERS, Agricultural Economic Report No. 391 (November).

_____. 1977. *Fresh Deciduous Fruit Reports.* Washington, DC: Foreign Agricultural Economics, Washington State University.

_____. 1979. *Apple Production by Major Varieties for Selected Countries.* Washington, DC: Horticultural and Tropical Products Division, Foreign Agriculture Service.

_____. 1991. *Fruit and Tree Nuts. Situation and Outlook Report.* Washington, DC: Economic Research Service, TFS-257 (March).

_____. 1991. *Fruit and Tree Nuts. Situation and Outlook Yearbook.* Washington, DC: Economics Research Service, TFS-258 (August).

_____. 1983. *Nationwide Food Consumption Survey, 1977-78.* Washington, DC.

_____. *Noncitrus Fruits and Nuts.* Washington, DC: National Agricultural Statistics Service, annual.

U.S. Department of Commerce. 1991. *Advance Monthly Retail Sales, December 1990.* Washington, DC: Economics and Statistics Administration, Bureau of Census (January).

U.S. International Trade Commission. 1986. *Apple Juice.* Washington, DC: Report to the President on Investigation No. TA-201-59, publication 1861 (June).

Urban, Francis and Michael Trueblood. 1990. *World Population by Country and Region, 1950-2050.* Washington, DC: U.S. Department of Agriculture, Economic Research Service.

Vidyashankara, Satyanarayana, and Wesley W. Wilson. 1989. *World Trade in Apples: 1962-1987.* Pullman, WA: Washington State University, IMPACT Center, Information Series, No. 31.

Zind, Tom. 1987. "Fresh Trends." *The Packer,* Focus 1986-87, 33-74.

Index

T - #0023 - 230425 - C0 - 229/152/14 [16] - CB - 9781560220411 - Gloss Lamination